D1152363

nec3

NORWICH CITY COLLEGE LIBRARY

Stock No.	223543		
Class	692.8 NEC		
Cat.	O/S	Proc.	3WKU

Engineering and Construction Contract

This contract should be used for the appointment of a contractor for engineering and construction work, including any level of design responsibility

An NEC document

June 2005
(with amendments June 2006)

OGC endorsement of NEC3

OGC advises public sector procurers that the form of contract used has to be selected according to the objectives of the project, aiming to satisfy the *Achieving Excellence in Construction* (AEC) principles.

This edition of the NEC (NEC3) complies fully with the AEC principles. OGC recommends the use of NEC3 by public sector construction procurers on their construction projects.

OGC
Office of Government Commerce

223 543

NEC is a division of Thomas Telford Ltd, which is a wholly owned subsidiary of the Institution of Civil Engineers (ICE), the owner and developer of the NEC.

The NEC is a family of standard contracts, each of which has these characteristics:

- Its use stimulates good management of the relationship between the two parties to the contract and, hence, of the work included in the contract.
- It can be used in a wide variety of commercial situations, for a wide variety of types of work and in any location.
- It is a clear and simple document – using language and a structure which are straightforward and easily understood.

NEC3 Engineering and Construction Contract is one of the NEC family and is consistent with all other NEC3 documents. Also available are the Engineering and Construction Contract Guidance Notes, Flow Charts and Options A, B, C, D, E and F.

ISBN (complete box set) 0 7277 3382 6
ISBN (this document) 0 7277 3359 1
ISBN (Engineering and Construction Contract Guidance Notes) 0 7277 3366 4
ISBN (Engineering and Construction Contract Flow Charts) 0 7277 3367 2
ISBN (Option A: Priced contract with activity schedule) 0 7277 3360 5
ISBN (Option B: Priced contract with bill of quantities) 0 7277 3361 3
ISBN (Option C: Target contract with activity schedule) 0 7277 3362 1
ISBN (Option D: Target contract with bill of quantities) 0 7277 3363 X
ISBN (Option E: Cost reimbursable contract) 0 7277 3364 8
ISBN (Option F: Management contract) 0 7277 3365 6

Consultative edition 1991
First edition 1993
Second edition 1995
Reprinted November 1995 with amendments
Reprinted with corrections May 1998
Third edition June 2005
Reprinted with amendments June 2006

Cover photo, Golden Jubilee Bridge, courtesy of City of Westminster

9 8 7 6 5 4 3

British Library Cataloguing in Publication Data for this publication is available from the British Library.

© Copyright nec 2005

All rights, including translation, reserved. The owner of this document may reproduce the Contract Data for the purpose of obtaining tenders, awarding and administering contracts. Except as permitted by the Copyright, Designs and Patents Act 1988, no part of this publication may be otherwise reproduced, stored in a retrieval system or transmitted in any form or by any means, electronic, mechanical, photocopying, recording or otherwise, without the prior written permission of the NEC Director, NEC, Thomas Telford Ltd, 1 Heron Quay, London E14 4JD.

Typeset by Academic + Technical, Bristol

Printed and bound in Great Britain by Bell & Bain Limited, Glasgow, UK

CONTENTS

Acknowledgements			iv
Schedule of Options			1
Amendments			2
Core clauses	1	General	3
	2	The *Contractor*'s main responsibilities	7
	3	Time	9
	4	Testing and Defects	11
	5	Payment	13
	6	Compensation events	15
	7	Title	20
	8	Risks and insurance	21
	9	Termination	24
Main Option clauses	A	Priced contract with activity schedule	27
	B	Priced contract with bill of quantities	29
	C	Target contract with activity schedule	31
	D	Target contract with bill of quantities	34
	E	Cost reimbursable contract	38
	F	Management contract	40
Dispute resolution	W	Option W1	42
		Option W2	45
Secondary Option clauses	X1	Price adjustment for inflation	48
	X2	Changes in the law	49
	X3	Multiple currencies	49
	X4	Parent company guarantee	49
	X5	Sectional Completion	49
	X6	Bonus for early Completion	50
	X7	Delay damages	50
	X12	Partnering	50
	X13	Performance bond	52
	X14	Advanced payment to the *Contractor*	52
	X15	Limitation of the *Contractor*'s liability for his design to reasonable skill and care	52
	X16	Retention	53
	X17	Low performance damages	53
	X18	Limitation of liability	53
	X20	Key Performance Indicators	54
	Y(UK)2	The Housing Grants, Construction and Regeneration Act 1996	55
	Y(UK)3	The Contracts (Rights of Third Parties) Act 1999	55
	Z	*Additional conditions of contract*	55
Note		Options X8 to X11, X19 and Y(UK)1 are not used	
Schedule of Cost Components			56
Shorter Schedule of Cost Components			59
Contract Data			61
Index			74

ACKNOWLEDGEMENTS

The NEC first edition was produced by the Institution of Civil Engineers through its NEC Working Group.

The original NEC was designed and drafted by Dr Martin Barnes then of Coopers and Lybrand with the assistance of Professor J. G. Perry then of the University of Birmingham, T. W. Weddell then of Travers Morgan Management, T. H. Nicholson, Consultant to the Institution of Civil Engineers, A. Norman then of the University of Manchester Institute of Science and Technology and P. A. Baird, then Corporate Contracts Consultant, Eskom, South Africa.

The second edition of the NEC documents for engineering and construction contracts was produced by the Institution of Civil Engineers through its NEC Panel.

The third edition of the NEC Engineering and Construction Contract was produced by the Institution of Civil Engineers through its NEC Panel. The Flow Charts were produced by John S. Gillespie with assistance from Tom Nicholson.

The members of the NEC Panel are:

P. Higgins, BSc, CEng, FICE, FCIArb (Chairman)
P. A. Baird, BSc, CEng, FICE, M(SA)ICE, MAPM
M. Barnes, BSc(Eng), PhD, FREng, FICE, FCIOB, CCMI, ACIArb, MBCS, FInstCES, FAPM
A. J. Bates, FRICS, MInstCES
A. J. M. Blackler, BA, LLB(Cantab), MCIArb
P. T. Cousins, BEng(Tech), DipArb, CEng, MICE, MCIArb, MCMI
L. T. Eames, BSc, FRICS, FCIOB
F. Forward, BA(Hons), DipArch, MSc(Const Law), RIBA, FCIArb
Professor J. G. Perry, MEng, PhD, CEng, FICE, MAPM
N. C. Shaw, FCIPS, CEng, MIMechE
T. W. Weddell, BSc, CEng, DIC, FICE, FIStructE, ACIArb

NEC Consultant:

R. A. Gerrard, BSc(Hons), MRICS, FCIArb, FInstCES

Secretariat:

A. Cole, LLB, LLM, BL
J. M. Hawkins, BA(Hons), MSc
F. N. Vernon (Technical Adviser), BSc, CEng, MICE

The Institution of Civil Engineers acknowledges the help in preparing the third edition given by many other people, in particular, by:

J. C. Broome, BEng
A. Else, BSc, CEng, FICE
C. Flook, Esq.
R. Lewendon, FICE, MIHT, MCIArb, MAPM
T. H. Nicholson, BSc, FICE
R. Patterson, BA, MA(Cantab), MBA, CEng, MICE
C. Reed, CEng, MA, MSc, FICE
D. Weeks, FRICS
S. Zarka, FRICS

SCHEDULE OF OPTIONS

The strategy for choosing the form of contract starts with a decision between six main Options, one of which must be chosen.

Option A	Priced contract with activity schedule
Option B	Priced contract with bill of quantities
Option C	Target contract with activity schedule
Option D	Target contract with bill of quantities
Option E	Cost reimbursable contract
Option F	Management contract
	One of the following dispute resolution Options must be selected to complete the chosen main Option.
Option W1	Dispute resolution procedure (used unless the United Kingdom Housing Grants, Construction and Regeneration Act 1996 applies).
Option W2	Dispute resolution procedure (used in the United Kingdom when the Housing Grants, Construction and Regeneration Act 1996 applies).
	The following secondary Options should then be considered. It is not necessary to use any of them. Any combination other than those stated may be used.
Option X1	Price adjustment for inflation (used only with Options A, B, C and D)
Option X2	Changes in the law
Option X3	Multiple currencies (used only with Options A and B)
Option X4	Parent company guarantee
Option X5	Sectional Completion
Option X6	Bonus for early Completion
Option X7	Delay damages
Option X12	Partnering
Option X13	Performance bond
Option X14	Advanced payment to the *Contractor*
Option X15	Limitation of the *Contractor*'s liability for his design to reasonable skill and care
Option X16	Retention (not used with Option F)
Option X17	Low performance damages
Option X18	Limitation of liability
Option X20	Key Performance Indicators (not used with Option X12)
	The following Options dealing with national legislation should be included if required.
Option Y(UK)2	The Housing Grants, Construction and Regeneration Act 1996
Option Y(UK)3	The Contracts (Rights of Third Parties) Act 1999
Option Z	*Additional conditions of contract*
Note	Options X8 to X11, X19 and Y(UK)1 are not used.

© copyright nec 2005

AMENDMENTS JUNE 2006

The following amendments have been made to the June 2005 edition.

Page	Clause	Line
10	32.1	4 deleted: 'and of notified early warning matters'
61	Contract Data Part one, 1.	3 added, '(with amendments June 2006)'
64	Contract Data Part one, 1.	23 deleted: 'If there are additional compensation events' and following 4 lines

 © copyright nec 2005

CORE CLAUSES

1 General

Actions **10**

10.1 The *Employer*, the *Contractor*, the *Project Manager* and the *Supervisor* shall act as stated in this contract and in a spirit of mutual trust and co-operation.

Identified and defined terms **11**

11.1 In these conditions of contract, terms identified in the Contract Data are in italics and defined terms have capital initials.

11.2 (1) The Accepted Programme is the programme identified in the Contract Data or is the latest programme accepted by the *Project Manager*. The latest programme accepted by the *Project Manager* supersedes previous Accepted Programmes.

(2) Completion is when the *Contractor* has

- done all the work which the Works Information states he is to do by the Completion Date and
- corrected notified Defects which would have prevented the *Employer* from using the *works* and Others from doing their work.

If the work which the *Contractor* is to do by the Completion Date is not stated in the Works Information, Completion is when the *Contractor* has done all the work necessary for the *Employer* to use the *works* and for Others to do their work.

(3) The Completion Date is the *completion date* unless later changed in accordance with this contract.

(4) The Contract Date is the date when this contract came into existence.

(5) A Defect is

- a part of the *works* which is not in accordance with the Works Information or
- a part of the *works* designed by the *Contractor* which is not in accordance with the applicable law or the *Contractor*'s design which the *Project Manager* has accepted.

(6) The Defects Certificate is either a list of Defects that the *Supervisor* has notified before the *defects date* which the *Contractor* has not corrected or, if there are no such Defects, a statement that there are none.

(7) Equipment is items provided by the *Contractor* and used by him to Provide the Works and which the Works Information does not require him to include in the *works*.

(8) The Fee is the sum of the amounts calculated by applying the *subcontracted fee percentage* to the Defined Cost of subcontracted work and the *direct fee percentage* to the Defined Cost of other work.

© copyright nec 2005

(9) A Key Date is the date by which work is to meet the Condition stated. The Key Date is the *key date* stated in the Contract Data and the Condition is the *condition* stated in the Contract Data unless later changed in accordance with this contract.

(10) Others are people or organisations who are not the *Employer*, the *Project Manager*, the *Supervisor*, the *Adjudicator*, the *Contractor* or any employee, Subcontractor or supplier of the *Contractor*.

(11) The Parties are the *Employer* and the *Contractor*.

(12) Plant and Materials are items intended to be included in the *works*.

(13) To Provide the Works means to do the work necessary to complete the *works* in accordance with this contract and all incidental work, services and actions which this contract requires.

(14) The Risk Register is a register of the risks which are listed in the Contract Data and the risks which the *Project Manager* or the *Contractor* has notified as an early warning matter. It includes a description of the risk and a description of the actions which are to be taken to avoid or reduce the risk.

(15) The Site is the area within the *boundaries of the site* and the volumes above and below it which are affected by work included in this contract.

(16) Site Information is information which

- describes the Site and its surroundings and
- is in the documents which the Contract Data states it is in.

(17) A Subcontractor is a person or organisation who has a contract with the *Contractor* to

- construct or install part of the *works*,
- provide a service necessary to Provide the Works or
- supply Plant and Materials which the person or organisation has wholly or partly designed specifically for the *works*.

(18) The Working Areas are those parts of the *working areas* which are

- necessary for Providing the Works and
- used only for work in this contract

unless later changed in accordance with this contract.

(19) Works Information is information which either

- specifies and describes the *works* or
- states any constraints on how the *Contractor* Provides the Works

and is either

- in the documents which the Contract Data states it is in or
- in an instruction given in accordance with this contract.

Interpretation and the law **12**

12.1 In this contract, except where the context shows otherwise, words in the singular also mean in the plural and the other way round and words in the masculine also mean in the feminine and neuter.

12.2 This contract is governed by the *law of the contract*.

12.3 No change to this contract, unless provided for by the *conditions of contract*, has effect unless it has been agreed, confirmed in writing and signed by the Parties.

12.4 This contract is the entire agreement between the Parties.

 © copyright nec 2005

Communications **13**

13.1 Each instruction, certificate, submission, proposal, record, acceptance, notification, reply and other communication which this contract requires is communicated in a form which can be read, copied and recorded. Writing is in the *language of this contract*.

13.2 A communication has effect when it is received at the last address notified by the recipient for receiving communications or, if none is notified, at the address of the recipient stated in the Contract Data.

13.3 If this contract requires the *Project Manager*, the *Supervisor* or the *Contractor* to reply to a communication, unless otherwise stated in this contract, he replies within the *period for reply*.

13.4 The *Project Manager* replies to a communication submitted or resubmitted to him by the *Contractor* for acceptance. If his reply is not acceptance, the *Project Manager* states his reasons and the *Contractor* resubmits the communication within the *period for reply* taking account of these reasons. A reason for withholding acceptance is that more information is needed in order to assess the *Contractor*'s submission fully.

13.5 The *Project Manager* may extend the *period for reply* to a communication if the *Project Manager* and the *Contractor* agree to the extension before the reply is due. The *Project Manager* notifies the *Contractor* of the extension which has been agreed.

13.6 The *Project Manager* issues his certificates to the *Employer* and the *Contractor*. The *Supervisor* issues his certificates to the *Project Manager* and the *Contractor*.

13.7 A notification which this contract requires is communicated separately from other communications.

13.8 The *Project Manager* may withhold acceptance of a submission by the *Contractor*. Withholding acceptance for a reason stated in this contract is not a compensation event.

The *Project Manager* and the *Supervisor* **14**

14.1 The *Project Manager*'s or the *Supervisor*'s acceptance of a communication from the *Contractor* or of his work does not change the *Contractor*'s responsibility to Provide the Works or his liability for his design.

14.2 The *Project Manager* and the *Supervisor*, after notifying the *Contractor*, may delegate any of their actions and may cancel any delegation. A reference to an action of the *Project Manager* or the *Supervisor* in this contract includes an action by his delegate.

14.3 The *Project Manager* may give an instruction to the *Contractor* which changes the Works Information or a Key Date.

14.4 The *Employer* may replace the *Project Manager* or the *Supervisor* after he has notified the *Contractor* of the name of the replacement.

Adding to the Working Areas **15**

15.1 The *Contractor* may submit a proposal for adding an area to the Working Areas to the *Project Manager* for acceptance. A reason for not accepting is that the proposed area is either not necessary for Providing the Works or used for work not in this contract.

Early warning **16**

16.1 The *Contractor* and the *Project Manager* give an early warning by notifying the other as soon as either becomes aware of any matter which could

- increase the total of the Prices,
- delay Completion,
- delay meeting a Key Date or
- impair the performance of the *works* in use.

© copyright nec 2005

The *Contractor* may give an early warning by notifying the *Project Manager* of any other matter which could increase his total cost. The *Project Manager* enters early warning matters in the Risk Register. Early warning of a matter for which a compensation event has previously been notified is not required.

16.2 Either the *Project Manager* or the *Contractor* may instruct the other to attend a risk reduction meeting. Each may instruct other people to attend if the other agrees.

16.3 At a risk reduction meeting, those who attend co-operate in

- making and considering proposals for how the effect of the registered risks can be avoided or reduced,
- seeking solutions that will bring advantage to all those who will be affected,
- deciding on the actions which will be taken and who, in accordance with this contract, will take them and
- deciding which risks have now been avoided or have passed and can be removed from the Risk Register.

16.4 The *Project Manager* revises the Risk Register to record the decisions made at each risk reduction meeting and issues the revised Risk Register to the *Contractor*. If a decision needs a change to the Works Information, the *Project Manager* instructs the change at the same time as he issues the revised Risk Register.

Ambiguities and inconsistencies **17**

17.1 The *Project Manager* or the *Contractor* notifies the other as soon as either becomes aware of an ambiguity or inconsistency in or between the documents which are part of this contract. The *Project Manager* gives an instruction resolving the ambiguity or inconsistency.

Illegal and impossible requirements **18**

18.1 The *Contractor* notifies the *Project Manager* as soon as he considers that the Works Information requires him to do anything which is illegal or impossible. If the *Project Manager* agrees, he gives an instruction to change the Works Information appropriately.

Prevention **19**

19.1 If an event occurs which

- stops the *Contractor* completing the *works* or
- stops the *Contractor* completing the *works* by the date shown on the Accepted Programme,

and which

- neither Party could prevent and
- an experienced contractor would have judged at the Contract Date to have such a small chance of occurring that it would have been unreasonable for him to have allowed for it,

the *Project Manager* gives an instruction to the *Contractor* stating how he is to deal with the event.

 © copyright nec 2005

2 The *Contractor*'s main responsibilities

Providing the Works **20**

20.1 The *Contractor* Provides the Works in accordance with the Works Information.

The *Contractor*'s design **21**

21.1 The *Contractor* designs the parts of the *works* which the Works Information states he is to design.

21.2 The *Contractor* submits the particulars of his design as the Works Information requires to the *Project Manager* for acceptance. A reason for not accepting the *Contractor*'s design is that it does not comply with either the Works Information or the applicable law.

The *Contractor* does not proceed with the relevant work until the *Project Manager* has accepted his design.

21.3 The *Contractor* may submit his design for acceptance in parts if the design of each part can be assessed fully.

Using the *Contractor*'s design **22**

22.1 The *Employer* may use and copy the *Contractor*'s design for any purpose connected with construction, use, alteration or demolition of the *works* unless otherwise stated in the Works Information and for other purposes as stated in the Works Information.

Design of Equipment **23**

23.1 The *Contractor* submits particulars of the design of an item of Equipment to the *Project Manager* for acceptance if the *Project Manager* instructs him to. A reason for not accepting is that the design of the item will not allow the *Contractor* to Provide the Works in accordance with

- the Works Information,
- the *Contractor*'s design which the *Project Manager* has accepted or
- the applicable law.

People **24**

24.1 The *Contractor* either employs each key person named to do the job stated in the Contract Data or employs a replacement person who has been accepted by the *Project Manager*. The *Contractor* submits the name, relevant qualifications and experience of a proposed replacement person to the *Project Manager* for acceptance. A reason for not accepting the person is that his relevant qualifications and experience are not as good as those of the person who is to be replaced.

24.2 The *Project Manager* may, having stated his reasons, instruct the *Contractor* to remove an employee. The *Contractor* then arranges that, after one day, the employee has no further connection with the work included in this contract.

Working with the *Employer* and Others **25**

25.1 The *Contractor* co-operates with Others in obtaining and providing information which they need in connection with the *works*. He co-operates with Others and shares the Working Areas with them as stated in the Works Information.

25.2 The *Employer* and the *Contractor* provide services and other things as stated in the Works Information. Any cost incurred by the *Employer* as a result of the *Contractor* not providing the services and other things which he is to provide is assessed by the *Project Manager* and paid by the *Contractor.*

© copyright nec 2005

25.3 If the *Project Manager* decides that the work does not meet the Condition stated for a Key Date by the date stated and, as a result, the *Employer* incurs additional cost either

- in carrying out work or
- by paying an additional amount to Others in carrying out work

on the same project, the additional cost which the *Employer* has paid or will incur is paid by the *Contractor*. The *Project Manager* assesses the additional cost within four weeks of the date when the Condition for the Key Date is met. The *Employer*'s right to recover the additional cost is his only right in these circumstances.

Subcontracting **26**

26.1 If the *Contractor* subcontracts work, he is responsible for Providing the Works as if he had not subcontracted. This contract applies as if a Subcontractor's employees and equipment were the *Contractor*'s.

26.2 The *Contractor* submits the name of each proposed Subcontractor to the *Project Manager* for acceptance. A reason for not accepting the Subcontractor is that his appointment will not allow the *Contractor* to Provide the Works. The *Contractor* does not appoint a proposed Subcontractor until the *Project Manager* has accepted him.

26.3 The *Contractor* submits the proposed conditions of contract for each subcontract to the *Project Manager* for acceptance unless

- an NEC contract is proposed or
- the *Project Manager* has agreed that no submission is required.

The *Contractor* does not appoint a Subcontractor on the proposed subcontract conditions submitted until the *Project Manager* has accepted them. A reason for not accepting them is that

- they will not allow the *Contractor* to Provide the Works or
- they do not include a statement that the parties to the subcontract shall act in a spirit of mutual trust and co-operation.

Other responsibilities **27**

27.1 The *Contractor* obtains approval of his design from Others where necessary.

27.2 The *Contractor* provides access to work being done and to Plant and Materials being stored for this contract for

- the *Project Manager*,
- the *Supervisor* and
- Others notified to him by the *Project Manager*.

27.3 The *Contractor* obeys an instruction which is in accordance with this contract and is given to him by the *Project Manager* or the *Supervisor*.

27.4 The *Contractor* acts in accordance with the health and safety requirements stated in the Works Information.

 © copyright nec 2005

3 Time

Starting, Completion and Key Dates **30**

30.1 The *Contractor* does not start work on the Site until the first *access date* and does the work so that Completion is on or before the Completion Date.

30.2 The *Project Manager* decides the date of Completion. The *Project Manager* certifies Completion within one week of Completion.

30.3 The *Contractor* does the work so that the Condition stated for each Key Date is met by the Key Date.

The programme **31**

31.1 If a programme is not identified in the Contract Data, the *Contractor* submits a first programme to the *Project Manager* for acceptance within the period stated in the Contract Data.

31.2 The *Contractor* shows on each programme which he submits for acceptance

- the *starting date*, *access dates*, Key Dates and Completion Date,
- planned Completion,
- the order and timing of the operations which the *Contractor* plans to do in order to Provide the Works,
- the order and timing of the work of the *Employer* and Others as last agreed with them by the *Contractor* or, if not so agreed, as stated in the Works Information,
- the dates when the *Contractor* plans to meet each Condition stated for the Key Dates and to complete other work needed to allow the *Employer* and Others to do their work,
- provisions for
  - float,
  - time risk allowances,
  - health and safety requirements and
  - the procedures set out in this contract,
- the dates when, in order to Provide the Works in accordance with his programme, the *Contractor* will need
  - access to a part of the Site if later than its *access date*,
  - acceptances,
  - Plant and Materials and other things to be provided by the *Employer* and
  - information from Others,
- for each operation, a statement of how the *Contractor* plans to do the work identifying the principal Equipment and other resources which he plans to use and
- other information which the Works Information requires the *Contractor* to show on a programme submitted for acceptance.

31.3 Within two weeks of the *Contractor* submitting a programme to him for acceptance, the *Project Manager* either accepts the programme or notifies the *Contractor* of his reasons for not accepting it. A reason for not accepting a programme is that

- the *Contractor*'s plans which it shows are not practicable,
- it does not show the information which this contract requires,
- it does not represent the *Contractor*'s plans realistically or
- it does not comply with the Works Information.

© copyright nec 2005

Revising the programme **32**

32.1 The *Contractor* shows on each revised programme

- the actual progress achieved on each operation and its effect upon the timing of the remaining work,
- the effects of implemented compensation events,
- how the *Contractor* plans to deal with any delays and to correct notified Defects and
- any other changes which the *Contractor* proposes to make to the Accepted Programme.

32.2 The *Contractor* submits a revised programme to the *Project Manager* for acceptance

- within the *period for reply* after the *Project Manager* has instructed him to,
- when the *Contractor* chooses to and, in any case,
- at no longer interval than the interval stated in the Contract Data from the *starting date* until Completion of the whole of the *works*.

Access to and use of the Site **33**

33.1 The *Employer* allows access to and use of each part of the Site to the *Contractor* which is necessary for the work included in this contract. Access and use is allowed on or before the later of its *access date* and the date for access shown on the Accepted Programme.

Instructions to stop or not to start work **34**

34.1 The *Project Manager* may instruct the *Contractor* to stop or not to start any work and may later instruct him that he may re-start or start it.

Take over **35**

35.1 The *Employer* need not take over the *works* before the Completion Date if it is stated in the Contract Data that he is not willing to do so. Otherwise the *Employer* takes over the *works* not later than two weeks after Completion.

35.2 The *Employer* may use any part of the *works* before Completion has been certified. If he does so, he takes over the part of the *works* when he begins to use it except if the use is

- for a reason stated in the Works Information or
- to suit the *Contractor*'s method of working.

35.3 The *Project Manager* certifies the date upon which the *Employer* takes over any part of the *works* and its extent within one week of the date.

Acceleration **36**

36.1 The *Project Manager* may instruct the *Contractor* to submit a quotation for an acceleration to achieve Completion before the Completion Date. The *Project Manager* states changes to the Key Dates to be included in the quotation. A quotation for an acceleration comprises proposed changes to the Prices and a revised programme showing the earlier Completion Date and the changed Key Dates. The *Contractor* submits details of his assessment with each quotation.

36.2 The *Contractor* submits a quotation or gives his reasons for not doing so within the *period for reply*.

 © copyright nec 2005

4 Testing and Defects

Tests and inspections **40**

40.1 The subclauses in this clause only apply to tests and inspections required by the Works Information or the applicable law.

40.2 The *Contractor* and the *Employer* provide materials, facilities and samples for tests and inspections as stated in the Works Information.

40.3 The *Contractor* and the *Supervisor* each notifies the other of each of his tests and inspections before it starts and afterwards notifies the other of its results. The *Contractor* notifies the *Supervisor* in time for a test or inspection to be arranged and done before doing work which would obstruct the test or inspection. The *Supervisor* may watch any test done by the *Contractor*.

40.4 If a test or inspection shows that any work has a Defect, the *Contractor* corrects the Defect and the test or inspection is repeated.

40.5 The *Supervisor* does his tests and inspections without causing unnecessary delay to the work or to a payment which is conditional upon a test or inspection being successful. A payment which is conditional upon a *Supervisor*'s test or inspection being successful becomes due at the later of the *defects date* and the end of the last *defect correction period* if

- the *Supervisor* has not done the test or inspection and
- the delay to the test or inspection is not the *Contractor*'s fault.

40.6 The *Project Manager* assesses the cost incurred by the *Employer* in repeating a test or inspection after a Defect is found. The *Contractor* pays the amount assessed.

Testing and inspection before delivery **41**

41.1 The *Contractor* does not bring to the Working Areas those Plant and Materials which the Works Information states are to be tested or inspected before delivery until the *Supervisor* has notified the *Contractor* that they have passed the test or inspection.

Searching for and notifying Defects **42**

42.1 Until the *defects date*, the *Supervisor* may instruct the *Contractor* to search for a Defect. He gives his reason for the search with his instruction. Searching may include

- uncovering, dismantling, re-covering and re-erecting work,
- providing facilities, materials and samples for tests and inspections done by the *Supervisor* and
- doing tests and inspections which the Works Information does not require.

42.2 Until the *defects date*, the *Supervisor* notifies the *Contractor* of each Defect as soon as he finds it and the *Contractor* notifies the *Supervisor* of each Defect as soon as he finds it.

Correcting Defects **43**

43.1 The *Contractor* corrects a Defect whether or not the *Supervisor* notifies him of it.

43.2 The *Contractor* corrects a notified Defect before the end of the *defect correction period*. The *defect correction period* begins at Completion for Defects notified before Completion and when the Defect is notified for other Defects.

© copyright nec 2005

43.3 The *Supervisor* issues the Defects Certificate at the later of the *defects date* and the end of the last *defect correction period*. The *Employer*'s rights in respect of a Defect which the *Supervisor* has not found or notified are not affected by the issue of the Defects Certificate.

43.4 The *Project Manager* arranges for the *Employer* to allow the *Contractor* access to and use of a part of the *works* which he has taken over if they are needed for correcting a Defect. In this case the *defect correction period* begins when the necessary access and use have been provided.

Accepting Defects 44

44.1 The *Contractor* and the *Project Manager* may each propose to the other that the Works Information should be changed so that a Defect does not have to be corrected.

44.2 If the *Contractor* and the *Project Manager* are prepared to consider the change, the *Contractor* submits a quotation for reduced Prices or an earlier Completion Date or both to the *Project Manager* for acceptance. If the *Project Manager* accepts the quotation, he gives an instruction to change the Works Information, the Prices and the Completion Date accordingly.

Uncorrected Defects 45

45.1 If the *Contractor* is given access in order to correct a notified Defect but he has not corrected it within its *defect correction period*, the *Project Manager* assesses the cost to the *Employer* of having the Defect corrected by other people and the *Contractor* pays this amount. The Works Information is treated as having been changed to accept the Defect.

45.2 If the *Contractor* is not given access in order to correct a notified Defect before the *defects date*, the *Project Manager* assesses the cost to the *Contractor* of correcting the Defect and the *Contractor* pays this amount. The Works Information is treated as having been changed to accept the Defect.

 © copyright nec 2005

5 Payment

Assessing the amount due **50**

50.1 The *Project Manager* assesses the amount due at each assessment date. The first assessment date is decided by the *Project Manager* to suit the procedures of the Parties and is not later than the *assessment interval* after the *starting date*. Later assessment dates occur

- at the end of each *assessment interval* until four weeks after the *Supervisor* issues the Defects Certificate and
- at Completion of the whole of the *works*.

50.2 The amount due is

- the Price for Work Done to Date,
- plus other amounts to be paid to the *Contractor*,
- less amounts to be paid by or retained from the *Contractor*.

Any tax which the law requires the *Employer* to pay to the *Contractor* is included in the amount due.

50.3 If no programme is identified in the Contract Data, one quarter of the Price for Work Done to Date is retained in assessments of the amount due until the *Contractor* has submitted a first programme to the *Project Manager* for acceptance showing the information which this contract requires.

50.4 In assessing the amount due, the *Project Manager* considers any application for payment the *Contractor* has submitted on or before the assessment date. The *Project Manager* gives the *Contractor* details of how the amount due has been assessed.

50.5 The *Project Manager* corrects any wrongly assessed amount due in a later payment certificate.

Payment **51**

51.1 The *Project Manager* certifies a payment within one week of each assessment date. The first payment is the amount due. Other payments are the change in the amount due since the last payment certificate. A payment is made by the *Contractor* to the *Employer* if the change reduces the amount due. Other payments are made by the *Employer* to the *Contractor*. Payments are in the *currency of this contract* unless otherwise stated in this contract.

51.2 Each certified payment is made within three weeks of the assessment date or, if a different period is stated in the Contract Data, within the period stated. If a certified payment is late, or if a payment is late because the *Project Manager* does not issue a certificate which he should issue, interest is paid on the late payment. Interest is assessed from the date by which the late payment should have been made until the date when the late payment is made, and is included in the first assessment after the late payment is made.

51.3 If an amount due is corrected in a later certificate either

- by the *Project Manager* in relation to a mistake or a compensation event or
- following a decision of the *Adjudicator* or the *tribunal*,

interest on the correcting amount is paid. Interest is assessed from the date when the incorrect amount was certified until the date when the correcting amount is certified and is included in the assessment which includes the correcting amount.

51.4 Interest is calculated on a daily basis at the *interest rate* and is compounded annually.

 © copyright nec 2005

Defined Cost 52

52.1 All the *Contractor*'s costs which are not included in the Defined Cost are treated as included in the Fee. Defined Cost includes only amounts calculated using rates and percentages stated in the Contract Data and other amounts at open market or competitively tendered prices with deductions for all discounts, rebates and taxes which can be recovered.

 © copyright nec 2005

6 Compensation events

Compensation events **60**

60.1 The following are compensation events.

(1) The *Project Manager* gives an instruction changing the Works Information except

- a change made in order to accept a Defect or
- a change to the Works Information provided by the *Contractor* for his design which is made either at his request or to comply with other Works Information provided by the *Employer*.

(2) The *Employer* does not allow access to and use of a part of the Site by the later of its *access date* and the date shown on the Accepted Programme.

(3) The *Employer* does not provide something which he is to provide by the date for providing it shown on the Accepted Programme.

(4) The *Project Manager* gives an instruction to stop or not to start any work or to change a Key Date.

(5) The *Employer* or Others

- do not work within the times shown on the Accepted Programme,
- do not work within the conditions stated in the Works Information or
- carry out work on the Site that is not stated in the Works Information.

(6) The *Project Manager* or the *Supervisor* does not reply to a communication from the *Contractor* within the period required by this contract.

(7) The *Project Manager* gives an instruction for dealing with an object of value or of historical or other interest found within the Site.

(8) The *Project Manager* or the *Supervisor* changes a decision which he has previously communicated to the *Contractor.*

(9) The *Project Manager* withholds an acceptance (other than acceptance of a quotation for acceleration or for not correcting a Defect) for a reason not stated in this contract.

(10) The *Supervisor* instructs the *Contractor* to search for a Defect and no Defect is found unless the search is needed only because the *Contractor* gave insufficient notice of doing work obstructing a required test or inspection.

(11) A test or inspection done by the *Supervisor* causes unnecessary delay.

(12) The *Contractor* encounters physical conditions which

- are within the Site,
- are not weather conditions and
- an experienced contractor would have judged at the Contract Date to have such a small chance of occurring that it would have been unreasonable for him to have allowed for them.

Only the difference between the physical conditions encountered and those for which it would have been reasonable to have allowed is taken into account in assessing a compensation event.

(13) A *weather measurement* is recorded

- within a calendar month,
- before the Completion Date for the whole of the *works* and
- at the place stated in the Contract Data

the value of which, by comparison with the *weather data*, is shown to occur on average less frequently than once in ten years.

 © copyright nec 2005

Only the difference between the *weather measurement* and the weather which the *weather data* show to occur on average less frequently than once in ten years is taken into account in assessing a compensation event.

(14) An event which is an *Employer*'s risk stated in this contract.

(15) The *Project Manager* certifies take over of a part of the *works* before both Completion and the Completion Date.

(16) The *Employer* does not provide materials, facilities and samples for tests and inspections as stated in the Works Information.

(17) The *Project Manager* notifies a correction to an assumption which he has stated about a compensation event.

(18) A breach of contract by the *Employer* which is not one of the other compensation events in this contract.

(19) An event which

- stops the *Contractor* completing the *works* or
- stops the *Contractor* completing the *works* by the date shown on the Accepted Programme,

and which

- neither Party could prevent,
- an experienced contractor would have judged at the Contract Date to have such a small chance of occurring that it would have been unreasonable for him to have allowed for it and
- is not one of the other compensation events stated in this contract.

60.2 In judging the physical conditions for the purpose of assessing a compensation event, the *Contractor* is assumed to have taken into account

- the Site Information,
- publicly available information referred to in the Site Information,
- information obtainable from a visual inspection of the Site and
- other information which an experienced contractor could reasonably be expected to have or to obtain.

60.3 If there is an ambiguity or inconsistency within the Site Information (including the information referred to in it), the *Contractor* is assumed to have taken into account the physical conditions more favourable to doing the work.

Notifying compensation events **61**

61.1 For compensation events which arise from the *Project Manager* or the *Supervisor* giving an instruction or changing an earlier decision, the *Project Manager* notifies the *Contractor* of the compensation event at the time of giving the instruction or changing the earlier decision. He also instructs the *Contractor* to submit quotations, unless the event arises from a fault of the *Contractor* or quotations have already been submitted. The *Contractor* puts the instruction or changed decision into effect.

61.2 The *Project Manager* may instruct the *Contractor* to submit quotations for a proposed instruction or a proposed changed decision. The *Contractor* does not put a proposed instruction or a proposed changed decision into effect.

61.3 The *Contractor* notifies the *Project Manager* of an event which has happened or which he expects to happen as a compensation event if

- the *Contractor* believes that the event is a compensation event and
- the *Project Manager* has not notified the event to the *Contractor*.

If the *Contractor* does not notify a compensation event within eight weeks of becoming aware of the event, he is not entitled to a change in the Prices, the Completion Date or a Key Date unless the *Project Manager* should have notified the event to the *Contractor* but did not.

 © copyright nec 2005

61.4 If the *Project Manager* decides that an event notified by the *Contractor*

- arises from a fault of the *Contractor*,
- has not happened and is not expected to happen,
- has no effect upon Defined Cost, Completion or meeting a Key Date or
- is not one of the compensation events stated in this contract

he notifies the *Contractor* of his decision that the Prices, the Completion Date and the Key Dates are not to be changed.

If the *Project Manager* decides otherwise, he notifies the *Contractor* accordingly and instructs him to submit quotations.

If the *Project Manager* does not notify his decision to the *Contractor* within either

- one week of the *Contractor*'s notification or
- a longer period to which the *Contractor* has agreed,

the *Contractor* may notify the *Project Manager* to this effect. A failure by the *Project Manager* to reply within two weeks of this notification is treated as acceptance by the *Project Manager* that the event is a compensation event and an instruction to submit quotations.

61.5 If the *Project Manager* decides that the *Contractor* did not give an early warning of the event which an experienced contractor could have given, he notifies this decision to the *Contractor* when he instructs him to submit quotations.

61.6 If the *Project Manager* decides that the effects of a compensation event are too uncertain to be forecast reasonably, he states assumptions about the event in his instruction to the *Contractor* to submit quotations. Assessment of the event is based on these assumptions. If any of them is later found to have been wrong, the *Project Manager* notifies a correction.

61.7 A compensation event is not notified after the *defects date*.

Quotations for compensation events **62**

62.1 After discussing with the *Contractor* different ways of dealing with the compensation event which are practicable, the *Project Manager* may instruct the *Contractor* to submit alternative quotations. The *Contractor* submits the required quotations to the *Project Manager* and may submit quotations for other methods of dealing with the compensation event which he considers practicable.

62.2 Quotations for compensation events comprise proposed changes to the Prices and any delay to the Completion Date and Key Dates assessed by the *Contractor*. The *Contractor* submits details of his assessment with each quotation. If the programme for remaining work is altered by the compensation event, the *Contractor* includes the alterations to the Accepted Programme in his quotation.

62.3 The *Contractor* submits quotations within three weeks of being instructed to do so by the *Project Manager*. The *Project Manager* replies within two weeks of the submission. His reply is

- an instruction to submit a revised quotation,
- an acceptance of a quotation,
- a notification that a proposed instruction will not be given or a proposed changed decision will not be made or
- a notification that he will be making his own assessment.

62.4 The *Project Manager* instructs the *Contractor* to submit a revised quotation only after explaining his reasons for doing so to the *Contractor*. The *Contractor* submits the revised quotation within three weeks of being instructed to do so.

 © copyright nec 2005

62.5 The *Project Manager* extends the time allowed for

- the *Contractor* to submit quotations for a compensation event and
- the *Project Manager* to reply to a quotation

if the *Project Manager* and the *Contractor* agree to the extension before the submission or reply is due. The *Project Manager* notifies the extension that has been agreed to the *Contractor*.

62.6 If the *Project Manager* does not reply to a quotation within the time allowed, the *Contractor* may notify the *Project Manager* to this effect. If the *Contractor* submitted more than one quotation for the compensation event, he states in his notification which quotation he proposes is to be accepted. If the *Project Manager* does not reply to the notification within two weeks, and unless the quotation is for a proposed instruction or a proposed changed decision, the *Contractor*'s notification is treated as acceptance of the quotation by the *Project Manager*.

Assessing compensation events **63**

63.1 The changes to the Prices are assessed as the effect of the compensation event upon

- the actual Defined Cost of the work already done,
- the forecast Defined Cost of the work not yet done and
- the resulting Fee.

The date when the *Project Manager* instructed or should have instructed the *Contractor* to submit quotations divides the work already done from the work not yet done.

63.2 If the effect of a compensation event is to reduce the total Defined Cost, the Prices are not reduced except as stated in this contract.

63.3 A delay to the Completion Date is assessed as the length of time that, due to the compensation event, planned Completion is later than planned Completion as shown on the Accepted Programme. A delay to a Key Date is assessed as the length of time that, due to the compensation event, the planned date when the Condition stated for a Key Date will be met is later than the date shown on the Accepted Programme.

63.4 The rights of the *Employer* and the *Contractor* to changes to the Prices, the Completion Date and the Key Dates are their only rights in respect of a compensation event.

63.5 If the *Project Manager* has notified the *Contractor* of his decision that the *Contractor* did not give an early warning of a compensation event which an experienced contractor could have given, the event is assessed as if the *Contractor* had given early warning.

63.6 Assessment of the effect of a compensation event includes risk allowances for cost and time for matters which have a significant chance of occurring and are at the *Contractor*'s risk under this contract.

63.7 Assessments are based upon the assumptions that the *Contractor* reacts competently and promptly to the compensation event, that any Defined Cost and time due to the event are reasonably incurred and that the Accepted Programme can be changed.

63.8 A compensation event which is an instruction to change the Works Information in order to resolve an ambiguity or inconsistency is assessed as if the Prices, the Completion Date and the Key Dates were for the interpretation most favourable to the Party which did not provide the Works Information.

63.9 If a change to the Works Information makes the description of the Condition for a Key Date incorrect, the *Project Manager* corrects the description. This correction is taken into account in assessing the compensation event for the change to the Works Information.

 © copyright nec 2005

The *Project Manager*'s assessments **64**

64.1 The *Project Manager* assesses a compensation event

- if the *Contractor* has not submitted a quotation and details of his assessment within the time allowed,
- if the *Project Manager* decides that the *Contractor* has not assessed the compensation event correctly in a quotation and he does not instruct the *Contractor* to submit a revised quotation,
- if, when the *Contractor* submits quotations for a compensation event, he has not submitted a programme or alterations to a programme which this contract requires him to submit or
- if, when the *Contractor* submits quotations for a compensation event, the *Project Manager* has not accepted the *Contractor*'s latest programme for one of the reasons stated in this contract.

64.2 The *Project Manager* assesses a compensation event using his own assessment of the programme for the remaining work if

- there is no Accepted Programme or
- the *Contractor* has not submitted a programme or alterations to a programme for acceptance as required by this contract.

64.3 The *Project Manager* notifies the *Contractor* of his assessment of a compensation event and gives him details of it within the period allowed for the *Contractor*'s submission of his quotation for the same event. This period starts when the need for the *Project Manager*'s assessment becomes apparent.

64.4 If the *Project Manager* does not assess a compensation event within the time allowed, the *Contractor* may notify the *Project Manager* to this effect. If the *Contractor* submitted more than one quotation for the compensation event, he states in his notification which quotation he proposes is to be accepted. If the *Project Manager* does not reply within two weeks of this notification the notification is treated as acceptance of the *Contractor's* quotation by the *Project Manager*.

Implementing compensation events **65**

65.1 A compensation event is implemented when

- the *Project Manager* notifies his acceptance of the *Contractor*'s quotation,
- the *Project Manager* notifies the *Contractor* of his own assessment or
- a *Contractor*'s quotation is treated as having been accepted by the *Project Manager*.

65.2 The assessment of a compensation event is not revised if a forecast upon which it is based is shown by later recorded information to have been wrong.

© copyright nec 2005

7 Title

The *Employer*'s title to Plant and Materials **70**

70.1 Whatever title the *Contractor* has to Plant and Materials which is outside the Working Areas passes to the *Employer* if the *Supervisor* has marked it as for this contract.

70.2 Whatever title the *Contractor* has to Plant and Materials passes to the *Employer* if it has been brought within the Working Areas. The title to Plant and Materials passes back to the *Contractor* if it is removed from the Working Areas with the *Project Manager*'s permission.

Marking Equipment, Plant and Materials outside the Working Areas **71**

71.1 The *Supervisor* marks Equipment, Plant and Materials which are outside the Working Areas if

- this contract identifies them for payment and
- the *Contractor* has prepared them for marking as the Works Information requires.

Removing Equipment **72**

72.1 The *Contractor* removes Equipment from the Site when it is no longer needed unless the *Project Manager* allows it to be left in the *works*.

Objects and materials within the Site **73**

73.1 The *Contractor* has no title to an object of value or of historical or other interest within the Site. The *Contractor* notifies the *Project Manager* when such an object is found and the *Project Manager* instructs the *Contractor* how to deal with it. The *Contractor* does not move the object without instructions.

73.2 The *Contractor* has title to materials from excavation and demolition only as stated in the Works Information.

 © copyright nec 2005

8 Risks and insurance

Employer's risks **80**

80.1 The following are *Employer*'s risks.

- Claims, proceedings, compensation and costs payable which are due to
  - use or occupation of the Site by the *works* or for the purpose of the *works* which is the unavoidable result of the *works*,
  - negligence, breach of statutory duty or interference with any legal right by the *Employer* or by any person employed by or contracted to him except the *Contractor* or
  - a fault of the *Employer* or a fault in his design.
- Loss of or damage to Plant and Materials supplied to the *Contractor* by the *Employer*, or by Others on the *Employer*'s behalf, until the *Contractor* has received and accepted them.
- Loss of or damage to the *works*, Plant and Materials due to
  - war, civil war, rebellion, revolution, insurrection, military or usurped power,
  - strikes, riots and civil commotion not confined to the *Contractor*'s employees or
  - radioactive contamination.
- Loss of or wear or damage to the parts of the *works* taken over by the *Employer*, except loss, wear or damage occurring before the issue of the Defects Certificate which is due to
  - a Defect which existed at take over,
  - an event occurring before take over which was not itself an *Employer*'s risk or
  - the activities of the *Contractor* on the Site after take over.
- Loss of or wear or damage to the *works* and any Equipment, Plant and Materials retained on the Site by the *Employer* after a termination, except loss, wear or damage due to the activities of the *Contractor* on the Site after the termination.
- Additional *Employer*'s risks stated in the Contract Data.

The *Contractor*'s risks **81**

81.1 From the *starting date* until the Defects Certificate has been issued, the risks which are not carried by the *Employer* are carried by the *Contractor*.

Repairs **82**

82.1 Until the Defects Certificate has been issued and unless otherwise instructed by the *Project Manager*, the *Contractor* promptly replaces loss of and repairs damage to the *works*, Plant and Materials.

Indemnity **83**

83.1 Each Party indemnifies the other against claims, proceedings, compensation and costs due to an event which is at his risk.

83.2 The liability of each Party to indemnify the other is reduced if events at the other Party's risk contributed to the claims, proceedings, compensation and costs. The reduction is in proportion to the extent that events which were at the other Party's risk contributed, taking into account each Party's responsibilities under this contract.

© copyright nec 2005

Insurance cover **84**

84.1 The *Contractor* provides the insurances stated in the Insurance Table except any insurance which the *Employer* is to provide as stated in the Contract Data. The *Contractor* provides additional insurances as stated in the Contract Data.

84.2 The insurances are in the joint names of the Parties and provide cover for events which are at the *Contractor*'s risk from the *starting date* until the Defects Certificate or a termination certificate has been issued.

INSURANCE TABLE

Insurance against	Minimum amount of cover or minimum limit of indemnity
Loss of or damage to the *works*, Plant and Materials	The replacement cost, including the amount stated in the Contract Data for the replacement of any Plant and Materials provided by the *Employer*
Loss of or damage to Equipment	The replacement cost
Liability for loss of or damage to property (except the *works*, Plant and Materials and Equipment) and liability for bodily injury to or death of a person (not an employee of the *Contractor*) caused by activity in connection with this contract	The amount stated in the Contract Data for any one event with cross liability so that the insurance applies to the Parties separately
Liability for death of or bodily injury to employees of the *Contractor* arising out of and in the course of their employment in connection with this contract	The greater of the amount required by the applicable law and the amount stated in the Contract Data for any one event

Insurance policies **85**

85.1 Before the *starting date* and on each renewal of the insurance policy until the *defects date*, the *Contractor* submits to the *Project Manager* for acceptance certificates which state that the insurance required by this contract is in force. The certificates are signed by the *Contractor*'s insurer or insurance broker. A reason for not accepting the certificates is that they do not comply with this contract.

85.2 Insurance policies include a waiver by the insurers of their subrogation rights against directors and other employees of every insured except where there is fraud.

85.3 The Parties comply with the terms and conditions of the insurance policies.

85.4 Any amount not recovered from an insurer is borne by the *Employer* for events which are at his risk and by the *Contractor* for events which are at his risk.

If the *Contractor* does not insure **86**

86.1 The *Employer* may insure a risk which this contract requires the *Contractor* to insure if the *Contractor* does not submit a required certificate. The cost of this insurance to the *Employer* is paid by the *Contractor*.

Insurance by the *Employer* **87**

87.1 The *Project Manager* submits policies and certificates for insurances provided by the *Employer* to the *Contractor* for acceptance before the *starting date* and afterwards as the *Contractor* instructs. The *Contractor* accepts the policies and certificates if they comply with this contract.

 © copyright nec 2005

87.2 The *Contractor*'s acceptance of an insurance policy or certificate provided by the *Employer* does not change the responsibility of the *Employer* to provide the insurances stated in the Contract Data.

87.3 The *Contractor* may insure a risk which this contract requires the *Employer* to insure if the *Employer* does not submit a required policy or certificate. The cost of this insurance to the *Contractor* is paid by the *Employer*.

© copyright nec 2005

9 Termination

Termination 90

90.1 If either Party wishes to terminate the *Contractor*'s obligation to Provide the Works he notifies the *Project Manager* and the other Party giving details of his reason for terminating. The *Project Manager* issues a termination certificate to both Parties promptly if the reason complies with this contract.

90.2 The *Contractor* may terminate only for a reason identified in the Termination Table. The *Employer* may terminate for any reason. The procedures followed and the amounts due on termination are in accordance with the Termination Table.

TERMINATION TABLE

Terminating Party	Reason	Procedure	Amount due
The *Employer*	A reason other than R1–R21	P1 and P2	A1, A2 and A4
	R1–R15 or R18	P1, P2 and P3	A1 and A3
	R17 or R20	P1 and P3	A1 and A2
	R21	P1 and P4	A1 and A2
The *Contractor*	R1–R10, R16 or R19	P1 and P4	A1, A2 and A4
	R17 or R20	P1 and P4	A1 and A2

90.3 The procedures for termination are implemented immediately after the *Project Manager* has issued a termination certificate.

90.4 Within thirteen weeks of termination, the *Project Manager* certifies a final payment to or from the *Contractor* which is the *Project Manager*'s assessment of the amount due on termination less the total of previous payments. Payment is made within three weeks of the *Project Manager*'s certificate.

90.5 After a termination certificate has been issued, the *Contractor* does no further work necessary to Provide the Works.

Reasons for termination 91

91.1 Either Party may terminate if the other Party has done one of the following or its equivalent.

- If the other Party is an individual and has
  - presented his petition for bankruptcy (R1),
  - had a bankruptcy order made against him (R2),
  - had a receiver appointed over his assets (R3) or
  - made an arrangement with his creditors (R4).
- If the other Party is a company or partnership and has
  - had a winding-up order made against it (R5),
  - had a provisional liquidator appointed to it (R6),
  - passed a resolution for winding-up (other than in order to amalgamate or reconstruct) (R7),
  - had an administration order made against it (R8),
  - had a receiver, receiver and manager, or administrative receiver appointed over the whole or a substantial part of its undertaking or assets (R9) or
  - made an arrangement with its creditors (R10).

 © copyright nec 2005

91.2 The *Employer* may terminate if the *Project Manager* has notified that the *Contractor* has defaulted in one of the following ways and not put the default right within four weeks of the notification.

- Substantially failed to comply with his obligations (R11).
- Not provided a bond or guarantee which this contract requires (R12).
- Appointed a Subcontractor for substantial work before the *Project Manager* has accepted the Subcontractor (R13).

91.3 The *Employer* may terminate if the *Project Manager* has notified that the *Contractor* has defaulted in one of the following ways and not stopped defaulting within four weeks of the notification.

- Substantially hindered the *Employer* or Others (R14).
- Substantially broken a health or safety regulation (R15).

91.4 The *Contractor* may terminate if the *Employer* has not paid an amount certified by the *Project Manager* within thirteen weeks of the date of the certificate (R16).

91.5 Either Party may terminate if the Parties have been released under the law from further performance of the whole of this contract (R17).

91.6 If the *Project Manager* has instructed the *Contractor* to stop or not to start any substantial work or all work and an instruction allowing the work to re-start or start has not been given within thirteen weeks,

- the *Employer* may terminate if the instruction was due to a default by the *Contractor* (R18),
- the *Contractor* may terminate if the instruction was due to a default by the *Employer* (R19) and
- either Party may terminate if the instruction was due to any other reason (R20).

91.7 The *Employer* may terminate if an event occurs which

- stops the *Contractor* completing the *works* or
- stops the *Contractor* completing the *works* by the date shown on the Accepted Programme and is forecast to delay Completion by more than 13 weeks,

and which

- neither Party could prevent and
- an experienced contractor would have judged at the Contract Date to have such a small chance of occurring that it would have been unreasonable for him to have allowed for it (R21).

Procedures on termination 92

92.1 On termination, the *Employer* may complete the *works* and may use any Plant and Materials to which he has title (P1).

92.2 The procedure on termination also includes one or more of the following as set out in the Termination Table.

P2 The *Employer* may instruct the *Contractor* to leave the Site, remove any Equipment, Plant and Materials from the Site and assign the benefit of any subcontract or other contract related to performance of this contract to the *Employer*.

P3 The *Employer* may use any Equipment to which the *Contractor* has title to complete the *works*. The *Contractor* promptly removes the Equipment from Site when the *Project Manager* notifies him that the *Employer* no longer requires it to complete the *works*.

P4 The *Contractor* leaves the Working Areas and removes the Equipment.

© copyright nec 2005

Payment on termination **93**

93.1 The amount due on termination includes (A1)

- an amount due assessed as for normal payments,
- the Defined Cost for Plant and Materials
  - within the Working Areas or
  - to which the *Employer* has title and of which the *Contractor* has to accept delivery,
- other Defined Cost reasonably incurred in expectation of completing the whole of the *works*,
- any amounts retained by the *Employer* and
- a deduction of any un-repaid balance of an advanced payment.

93.2 The amount due on termination also includes one or more of the following as set out in the Termination Table.

A2 The forecast Defined Cost of removing the Equipment.

A3 A deduction of the forecast of the additional cost to the *Employer* of completing the whole of the *works*.

A4 The *direct fee percentage* applied to

- for Options A, B, C and D, any excess of the total of the Prices at the Contract Date over the Price for Work Done to Date or
- for Options E and F, any excess of the first forecast of the Defined Cost for the *works* over the Price for Work Done to Date less the Fee.

 © copyright nec 2005

MAIN OPTION CLAUSES

Option A: Priced contract with activity schedule

Identified and defined terms **11**

11.2 (20) The Activity Schedule is the *activity schedule* unless later changed in accordance with this contract.

(22) Defined Cost is the cost of the components in the Shorter Schedule of Cost Components whether work is subcontracted or not excluding the cost of preparing quotations for compensation events.

(27) The Price for Work Done to Date is the total of the Prices for

- each group of completed activities and
- each completed activity which is not in a group.

A completed activity is one which is without Defects which would either delay or be covered by immediately following work.

(30) The Prices are the lump sum prices for each of the activities on the Activity Schedule unless later changed in accordance with this contract.

The programme **31**

31.4 The *Contractor* provides information which shows how each activity on the Activity Schedule relates to the operations on each programme which he submits for acceptance.

Acceleration **36**

36.3 When the *Project Manager* accepts a quotation for an acceleration, he changes the Prices, the Completion Date and the Key Dates accordingly and accepts the revised programme.

The Activity Schedule **54**

54.1 Information in the Activity Schedule is not Works Information or Site Information.

54.2 If the *Contractor* changes a planned method of working at his discretion so that the activities on the Activity Schedule do not relate to the operations on the Accepted Programme, he submits a revision of the Activity Schedule to the *Project Manager* for acceptance.

54.3 A reason for not accepting a revision of the Activity Schedule is that

- it does not comply with the Accepted Programme,
- any changed Prices are not reasonably distributed between the activities or
- the total of the Prices is changed.

Assessing compensation events **63**

63.10 If the effect of a compensation event is to reduce the total Defined Cost and the event is

- a change to the Works Information or
- a correction of an assumption stated by the *Project Manager* for assessing an earlier compensation event,

the Prices are reduced.

63.12 Assessments for changed Prices for compensation events are in the form of changes to the Activity Schedule.

© copyright nec 2005

63.14 If the *Project Manager* and the *Contractor* agree, rates and lump sums may be used to assess a compensation event instead of Defined Cost.

Implementing compensation events **65**

65.4 The changes to the Prices, the Completion Date and the Key Dates are included in the notification implementing a compensation event.

Payment on termination **93**

93.3 The amount due on termination is assessed without taking grouping of activities into account.

 © copyright nec 2005

Option B: Priced contract with bill of quantities

Identified and defined terms **11**

11.2 (21) The Bill of Quantities is the *bill of quantities* as changed in accordance with this contract to accommodate implemented compensation events and for accepted quotations for acceleration.

(22) Defined Cost is the cost of the components in the Shorter Schedule of Cost Components whether work is subcontracted or not excluding the cost of preparing quotations for compensation events.

(28) The Price for Work Done to Date is the total of

- the quantity of the work which the *Contractor* has completed for each item in the Bill of Quantities multiplied by the rate and
- a proportion of each lump sum which is the proportion of the work covered by the item which the *Contractor* has completed.

Completed work is work without Defects which would either delay or be covered by immediately following work.

(31) The Prices are the lump sums and the amounts obtained by multiplying the rates by the quantities for the items in the Bill of Quantities.

Acceleration **36**

36.3 When the *Project Manager* accepts a quotation for an acceleration, he changes the Prices, the Completion Date and the Key Dates accordingly and accepts the revised programme.

The Bill of Quantities **55**

55.1 Information in the Bill of Quantities is not Works Information or Site Information.

Compensation events **60**

60.4 A difference between the final total quantity of work done and the quantity stated for an item in the Bill of Quantities is a compensation event if

- the difference does not result from a change to the Works Information,
- the difference causes the Defined Cost per unit of quantity to change and
- the rate in the Bill of Quantities for the item multiplied by the final total quantity of work done is more than 0.5% of the total of the Prices at the Contract Date.

If the Defined Cost per unit of quantity is reduced, the affected rate is reduced.

60.5 A difference between the final total quantity of work done and the quantity for an item stated in the Bill of Quantities which delays Completion or the meeting of the Condition stated for a Key Date is a compensation event.

60.6 The *Project Manager* corrects mistakes in the Bill of Quantities which are departures from the rules for item descriptions and for division of the work into items in the *method of measurement* or are due to ambiguities or inconsistencies. Each such correction is a compensation event which may lead to reduced Prices.

60.7 In assessing a compensation event which results from a correction of an inconsistency between the Bill of Quantities and another document, the *Contractor* is assumed to have taken the Bill of Quantities as correct.

 © copyright nec 2005

Assessing compensation events **63**

63.10 If the effect of a compensation event is to reduce the total Defined Cost and the event is

- a change to the Works Information or
- a correction of an assumption stated by the *Project Manager* for assessing an earlier compensation event,

the Prices are reduced.

63.13 Assessments for changed Prices for compensation events are in the form of changes to the Bill of Quantities.

- For the whole or a part of a compensation event for work not yet done and for which there is an item in the Bill of Quantities, the changes are
  - a changed rate,
  - a changed quantity or
  - a changed lump sum.
- For the whole or a part of a compensation event for work not yet done and for which there is no item in the Bill of Quantities, the change is a new priced item which, unless the *Project Manager* and the *Contractor* agree otherwise, is compiled in accordance with the *method of measurement.*
- For the whole or a part of a compensation event for work already done, the change is a new lump sum item.

If the *Project Manager* and the *Contractor* agree, rates and lump sums may be used to assess a compensation event instead of Defined Cost.

Implementing compensation events **65**

65.4 The changes to the Prices, the Completion Date and the Key Dates are included in the notification implementing a compensation event.

 © copyright nec 2005

Option C: Target contract with activity schedule

Identified and defined terms **11**

11.2 (20) The Activity Schedule is the *activity schedule* unless later changed in accordance with this contract.

(23) Defined Cost is

- the amount of payments due to Subcontractors for work which is subcontracted without taking account of amounts deducted for
  - retention,
  - payment to the *Employer* as a result of the Subcontractor failing to meet a Key Date,
  - the correction of Defects after Completion,
  - payments to Others and
  - the supply of equipment, supplies and services included in the charge for overhead cost within the Working Areas in this contract

and

- the cost of components in the Schedule of Cost Components for other work

less Disallowed Cost.

(25) Disallowed Cost is cost which the *Project Manager* decides

- is not justified by the *Contractor*'s accounts and records,
- should not have been paid to a Subcontractor or supplier in accordance with his contract,
- was incurred only because the *Contractor* did not
  - follow an acceptance or procurement procedure stated in the Works Information or
  - give an early warning which this contract required him to give

and the cost of

- correcting Defects after Completion,
- correcting Defects caused by the *Contractor* not complying with a constraint on how he is to Provide the Works stated in the Works Information,
- Plant and Materials not used to Provide the Works (after allowing for reasonable wastage) unless resulting from a change to the Works Information,
- resources not used to Provide the Works (after allowing for reasonable availability and utilisation) or not taken away from the Working Areas when the *Project Manager* requested and
- preparation for and conduct of an adjudication or proceedings of the *tribunal*.

(29) The Price for Work Done to Date is the total Defined Cost which the *Project Manager* forecasts will have been paid by the *Contractor* before the next assessment date plus the Fee.

(30) The Prices are the lump sum prices for each of the activities on the Activity Schedule unless later changed in accordance with this contract.

Providing the Works **20**

20.3 The *Contractor* advises the *Project Manager* on the practical implications of the design of the *works* and on subcontracting arrangements.

© copyright nec 2005

20.4 The *Contractor* prepares forecasts of the total Defined Cost for the whole of the *works* in consultation with the *Project Manager* and submits them to the *Project Manager*. Forecasts are prepared at the intervals stated in the Contract Data from the *starting date* until Completion of the whole of the *works*. An explanation of the changes made since the previous forecast is submitted with each forecast.

Subcontracting 26

26.4 The *Contractor* submits the proposed contract data for each subcontract for acceptance to the *Project Manager* if

- an NEC contract is proposed and
- the *Project Manager* instructs the *Contractor* to make the submission.

A reason for not accepting the proposed contract data is that its use will not allow the *Contractor* to Provide the Works.

The programme 31

31.4 The *Contractor* provides information which shows how each activity on the Activity Schedule relates to the operations on each programme which he submits for acceptance.

Acceleration 36

36.3 When the *Project Manager* accepts a quotation for an acceleration, he changes the Prices, the Completion Date and the Key Dates accordingly and accepts the revised programme.

Tests and inspections 40

40.7 When the *Project Manager* assesses the cost incurred by the *Employer* in repeating a test or inspection after a Defect is found, the *Project Manager* does not include the *Contractor*'s cost of carrying out the repeat test or inspection.

Assessing the amount due 50

50.6 Payments of Defined Cost made by the *Contractor* in a currency other than the *currency of this contract* are included in the amount due as payments to be made to him in the same currency. Such payments are converted to the *currency of this contract* in order to calculate the Fee and any *Contractor*'s share using the *exchange rates*.

Defined Cost 52

52.2 The *Contractor* keeps these records

- accounts of payments of Defined Cost,
- proof that the payments have been made,
- communications about and assessments of compensation events for Subcontractors and
- other records as stated in the Works Information.

52.3 The *Contractor* allows the *Project Manager* to inspect at any time within working hours the accounts and records which he is required to keep.

The *Contractor*'s share 53

53.1 The *Project Manager* assesses the *Contractor*'s share of the difference between the total of the Prices and the Price for Work Done to Date. The difference is divided into increments falling within each of the *share ranges*. The limits of a *share range* are the Price for Work Done to Date divided by the total of the Prices, expressed as a percentage. The *Contractor*'s share equals the sum of the products of the increment within each *share range* and the corresponding *Contractor's share percentage*.

53.2 If the Price for Work Done to Date is less than the total of the Prices, the *Contractor* is paid his share of the saving. If the Price for Work Done to Date is greater than the total of the Prices, the *Contractor* pays his share of the excess.

 © copyright nec 2005

53.3 The *Project Manager* makes a preliminary assessment of the *Contractor*'s share at Completion of the whole of the *works* using his forecasts of the final Price for Work Done to Date and the final total of the Prices. This share is included in the amount due following Completion of the whole of the *works*.

53.4 The *Project Manager* makes a final assessment of the *Contractor*'s share using the final Price for Work Done to Date and the final total of the Prices. This share is included in the final amount due.

The Activity Schedule **54**

54.1 Information in the Activity Schedule is not Works Information or Site Information.

54.2 If the *Contractor* changes a planned method of working at his discretion so that the activities on the Activity Schedule do not relate to the operations on the Accepted Programme, he submits a revision of the Activity Schedule to the *Project Manager* for acceptance.

54.3 A reason for not accepting a revision of the Activity Schedule is that

- it does not comply with the Accepted Programme,
- any changed Prices are not reasonably distributed between the activities or
- the total of the Prices is changed.

Assessing compensation events **63**

63.11 If the effect of a compensation event is to reduce the total Defined Cost and the event is

- a change to the Works Information, other than a change to the Works Information provided by the *Employer* which the *Contractor* proposed and the *Project Manager* has accepted or
- a correction of an assumption stated by the *Project Manager* for assessing an earlier compensation event,

the Prices are reduced.

63.12 Assessments for changed Prices for compensation events are in the form of changes to the Activity Schedule.

63.15 If the *Project Manager* and the *Contractor* agree, the *Contractor* assesses a compensation event using the Shorter Schedule of Cost Components. The *Project Manager* may make his own assessments using the Shorter Schedule of Cost Components.

Implementing compensation events **65**

65.4 The changes to the Prices, the Completion Date and the Key Dates are included in the notification implementing a compensation event.

Payment on termination **93**

93.4 If there is a termination, the *Project Manager* assesses the *Contractor*'s share after he has certified termination. His assessment uses, as the Price for Work Done to Date, the total of the Defined Cost which the *Contractor* has paid and which he is committed to pay for work done before termination. The assessment uses as the total of the Prices

- the lump sum price for each activity which has been completed and
- a proportion of the lump sum price for each incomplete activity which is the proportion of the work in the activity which has been completed.

93.6 The *Project Manager*'s assessment of the *Contractor*'s share is added to the amount due to the *Contractor* on termination if there has been a saving or deducted if there has been an excess.

© copyright nec 2005

Option D: Target contract with bill of quantities

Identified and defined terms **11** 11.2

(21) The Bill of Quantities is the *bill of quantities* as changed in accordance with this contract to accommodate implemented compensation events and for accepted quotations for acceleration.

(23) Defined Cost is

- the amount of payments due to Subcontractors for work which is subcontracted without taking account of amounts deducted for
  - retention,
  - payment to the *Employer* as a result of the Subcontractor failing to meet a Key Date,
  - the correction of Defects after Completion,
  - payments to Others and
  - the supply of equipment, supplies and services included in the charge for overhead cost within the Working Areas in this contract

and

- the cost of components in the Schedule of Cost Components for other work

less Disallowed Cost.

(25) Disallowed Cost is cost which the *Project Manager* decides

- is not justified by the *Contractor*'s accounts and records,
- should not have been paid to a Subcontractor or supplier in accordance with his contract,
- was incurred only because the *Contractor* did not
  - follow an acceptance or procurement procedure stated in the Works Information or
  - give an early warning which this contract required him to give

and the cost of

- correcting Defects after Completion,
- correcting Defects caused by the *Contractor* not complying with a constraint on how he is to Provide the Works stated in the Works Information,
- Plant and Materials not used to Provide the Works (after allowing for reasonable wastage) unless resulting from a change to the Works Information,
- resources not used to Provide the Works (after allowing for reasonable availability and utilisation) or not taken away from the Working Areas when the *Project Manager* requested and
- preparation for and conduct of an adjudication or proceedings of the *tribunal*.

(29) The Price for Work Done to Date is the total Defined Cost which the *Project Manager* forecasts will have been paid by the *Contractor* before the next assessment date plus the Fee.

(31) The Prices are the lump sums and the amounts obtained by multiplying the rates by the quantities for the items in the Bill of Quantities.

(33) The Total of the Prices is the total of

- the quantity of the work which the *Contractor* has completed for each item in the Bill of Quantities multiplied by the rate and
- a proportion of each lump sum which is the proportion of the work covered by the item which the *Contractor* has completed.

Completed work is work without Defects which would either delay or be covered by immediately following work.

 © copyright nec 2005

Providing the Works **20**

20.3 The *Contractor* advises the *Project Manager* on the practical implications of the design of the *works* and on subcontracting arrangements.

20.4 The *Contractor* prepares forecasts of the total Defined Cost for the whole of the *works* in consultation with the *Project Manager* and submits them to the *Project Manager*. Forecasts are prepared at the intervals stated in the Contract Data from the *starting date* until Completion of the whole of the *works*. An explanation of the changes made since the previous forecast is submitted with each forecast.

Subcontracting **26**

26.4 The *Contractor* submits the proposed contract data for each subcontract for acceptance to the *Project Manager* if

- an NEC contract is proposed and
- the *Project Manager* instructs the *Contractor* to make the submission.

A reason for not accepting the proposed contract data is that its use will not allow the *Contractor* to Provide the Works.

Acceleration **36**

36.3 When the *Project Manager* accepts a quotation for an acceleration, he changes the Prices, the Completion Date and the Key Dates accordingly and accepts the revised programme.

Tests and inspections **40**

40.7 When the *Project Manager* assesses the cost incurred by the *Employer* in repeating a test or inspection after a Defect is found, the *Project Manager* does not include the *Contractor*'s cost of carrying out the repeat test or inspection.

Assessing the amount due **50**

50.6 Payments of Defined Cost made by the *Contractor* in a currency other than the *currency of this contract* are included in the amount due as payments to be made to him in the same currency. Such payments are converted to the *currency of this contract* in order to calculate the Fee and any *Contractor*'s share using the *exchange rates.*

Defined Cost **52**

52.2 The *Contractor* keeps these records

- accounts of payments of Defined Cost,
- proof that the payments have been made,
- communications about and assessments of compensation events for Subcontractors and
- other records as stated in the Works Information.

52.3 The *Contractor* allows the *Project Manager* to inspect at any time within working hours the accounts and records which he is required to keep.

The *Contractor*'s share **53**

53.5 The *Project Manager* assesses the *Contractor*'s share of the difference between the Total of the Prices and the Price for Work Done to Date. The difference is divided into increments falling within each of the *share ranges*. The limits of a *share range* are the Price for Work Done to Date divided by the Total of the Prices, expressed as a percentage. The *Contractor*'s share equals the sum of the products of the increment within each *share range* and the corresponding *Contractor's share percentage.*

53.6 If the Price for Work Done to Date is less than the Total of the Prices, the *Contractor* is paid his share of the saving. If the Price for Work Done to Date is greater than the Total of the Prices, the *Contractor* pays his share of the excess.

© copyright nec 2005

53.7 The *Project Manager* makes a preliminary assessment of the *Contractor*'s share at Completion of the whole of the *works* using his forecasts of the final Price for Work Done to Date and the final Total of the Prices. This share is included in the amount due following Completion of the whole of the *works*.

53.8 The *Project Manager* makes a final assessment of the *Contractor*'s share using the final Price for Work Done to Date and the final Total of the Prices. This share is included in the final amount due.

The Bill of Quantities **55**

55.1 Information in the Bill of Quantities is not Works Information or Site Information.

Compensation events **60**

60.4 A difference between the final total quantity of work done and the quantity stated for an item in the Bill of Quantities is a compensation event if

- the difference does not result from a change to the Works Information,
- the difference causes the Defined Cost per unit of quantity to change and
- the rate in the Bill of Quantities for the item multiplied by the final total quantity of work done is more than 0.5% of the total of the Prices at the Contract Date.

If the Defined Cost per unit of quantity is reduced, the affected rate is reduced.

60.5 A difference between the final total quantity of work done and the quantity for an item stated in the Bill of Quantities which delays Completion or the meeting of the Condition stated for a Key Date is a compensation event.

60.6 The *Project Manager* corrects mistakes in the Bill of Quantities which are departures from the rules for item descriptions and for division of the work into items in the *method of measurement* or are due to ambiguities or inconsistencies. Each such correction is a compensation event which may lead to reduced Prices.

60.7 In assessing a compensation event which results from a correction of an inconsistency between the Bill of Quantities and another document, the *Contractor* is assumed to have taken the Bill of Quantities as correct.

Assessing compensation events **63**

63.11 If the effect of a compensation event is to reduce the total Defined Cost and the event is

- a change to the Works Information, other than a change to the Works Information provided by the *Employer* which the *Contractor* proposed and the *Project Manager* has accepted or
- a correction of an assumption stated by the *Project Manager* for assessing an earlier compensation event,

the Prices are reduced.

63.13 Assessments for changed Prices for compensation events are in the form of changes to the Bill of Quantities.

- For the whole or a part of a compensation event for work not yet done and for which there is an item in the Bill of Quantities, the changes are
  - a changed rate,
  - a changed quantity or
  - a changed lump sum.
- For the whole or a part of a compensation event for work not yet done and for which there is no item in the Bill of Quantities, the change is a new priced item which, unless the *Project Manager* and the *Contractor* agree otherwise, is compiled in accordance with the *method of measurement*.

 © copyright nec 2005

- For the whole or a part of a compensation event for work already done, the change is a new lump sum item.

If the *Project Manager* and the *Contractor* agree, rates and lump sums may be used to assess a compensation event instead of Defined Cost.

63.15 If the *Project Manager* and the *Contractor* agree, the *Contractor* assesses a compensation event using the Shorter Schedule of Cost Components. The *Project Manager* may make his own assessments using the Shorter Schedule of Cost Components.

Implementing compensation events **65**

65.4 The changes to the Prices, the Completion Date and the Key Dates are included in the notification implementing a compensation event.

Payment on termination **93**

93.5 If there is a termination, the *Project Manager* assesses the *Contractor*'s share after he has certified termination. His assessment uses, as the Price for Work Done to Date, the total of the Defined Cost which the *Contractor* has paid and which he is committed to pay for work done before termination.

93.6 The *Project Manager*'s assessment of the *Contractor*'s share is added to the amounts due to the *Contractor* on termination if there has been a saving or deducted if there has been an excess.

© copyright nec 2005

Option E: Cost reimbursable contract

Identified and defined terms **11**

11.2 (23) Defined Cost is

- the amount of payments due to Subcontractors for work which is subcontracted without taking account of amounts deducted for
  - retention,
  - payment to the *Employer* as a result of the Subcontractor failing to meet a Key Date,
  - the correction of Defects after Completion,
  - payments to Others and
  - the supply of equipment, supplies and services included in the charge for overhead cost within the Working Areas in this contract

and

- the cost of components in the Schedule of Cost Components for other work

less Disallowed Cost.

(25) Disallowed Cost is cost which the *Project Manager* decides

- is not justified by the *Contractor*'s accounts and records,
- should not have been paid to a Subcontractor or supplier in accordance with his contract,
- was incurred only because the *Contractor* did not
  - follow an acceptance or procurement procedure stated in the Works Information or
  - give an early warning which this contract required him to give

and the cost of

- correcting Defects after Completion,
- correcting Defects caused by the *Contractor* not complying with a constraint on how he is to Provide the Works stated in the Works Information,
- Plant and Materials not used to Provide the Works (after allowing for reasonable wastage) unless resulting from a change to the Works Information,
- resources not used to Provide the Works (after allowing for reasonable availability and utilisation) or not taken away from the Working Areas when the *Project Manager* requested and
- preparation for and conduct of an adjudication or proceedings of the *tribunal*.

(29) The Price for Work Done to Date is the total Defined Cost which the *Project Manager* forecasts will have been paid by the *Contractor* before the next assessment date plus the Fee.

(32) The Prices are the Defined Cost plus the Fee.

Providing the Works **20**

20.3 The *Contractor* advises the *Project Manager* on the practical implications of the design of the *works* and on subcontracting arrangements.

20.4 The *Contractor* prepares forecasts of the total Defined Cost for the whole of the *works* in consultation with the *Project Manager* and submits them to the *Project Manager*. Forecasts are prepared at the intervals stated in the Contract Data from the *starting date* until Completion of the whole of the *works*. An explanation of the changes made since the previous forecast is submitted with each forecast.

 © copyright nec 2005

Subcontracting **26**

26.4 The *Contractor* submits the proposed contract data for each subcontract for acceptance to the *Project Manager* if

- an NEC contract is proposed and
- the *Project Manager* instructs the *Contractor* to make the submission.

A reason for not accepting the proposed contract data is that its use will not allow the *Contractor* to Provide the Works.

Acceleration **36**

36.4 When the *Project Manager* accepts a quotation for an acceleration, he changes the Completion Date, the Key Dates and the forecast of the total Defined Cost of the whole of the *works* accordingly and accepts the revised programme.

Tests and inspections **40**

40.7 When the *Project Manager* assesses the cost incurred by the *Employer* in repeating a test or inspection after a Defect is found, the *Project Manager* does not include the *Contractor*'s cost of carrying out the repeat test or inspection.

Assessing the amount due **50**

50.7 Payments of Defined Cost made by the *Contractor* in a currency other than the *currency of this contract* are included in the amount due as payments to be made to him in the same currency. Such payments are converted to the *currency of this contract* in order to calculate the Fee using the *exchange rates.*

Defined Cost **52**

52.2 The *Contractor* keeps these records

- accounts of payments of Defined Cost,
- proof that the payments have been made,
- communications about and assessments of compensation events for Subcontractors and
- other records as stated in the Works Information.

52.3 The *Contractor* allows the *Project Manager* to inspect at any time within working hours the accounts and records which he is required to keep.

Assessing compensation events **63**

63.15 If the *Project Manager* and the *Contractor* agree, the *Contractor* assesses a compensation event using the Shorter Schedule of Cost Components. The *Project Manager* may make his own assessments using the Shorter Schedule of Cost Components.

Implementing compensation events **65**

65.3 The changes to the forecast amount of the Prices, the Completion Date and the Key Dates are included in the notification implementing a compensation event.

© copyright nec 2005

Option F: Management contract

Identified and defined terms **11**

11.2 (24) Defined Cost is

- the amount of payments due to Subcontractors for work which is subcontracted without taking account of amounts deducted for
  - retention,
  - payment to the *Employer* as a result of the Subcontractor failing to meet a Key Date,
  - the correction of Defects after Completion,
  - payments to Others,
  - the supply of equipment, supplies and services included in the charge for overhead cost within the Working Areas in this contract

and

- the *prices* for work done by the *Contractor* himself

less Disallowed Cost.

(26) Disallowed Cost is cost which the *Project Manager* decides

- is not justified by the accounts and records provided by the *Contractor*,
- should not have been paid to a Subcontractor or supplier in accordance with his contract,
- was incurred only because the *Contractor* did not
  - follow an acceptance or procurement procedure stated in the Works Information or
  - give an early warning which this contract required him to give or
- is a payment to a Subcontractor for
  - work which the Contract Data states that the *Contractor* will do himself or
  - the *Contractor*'s management.

(29) The Price for Work Done to Date is the total Defined Cost which the *Project Manager* forecasts will have been paid by the *Contractor* before the next assessment date plus the Fee.

(32) The Prices are the Defined Cost plus the Fee.

Providing the Works **20**

20.2 The *Contractor* manages the *Contractor*'s design, the provision of Site services and the construction and installation of the *works*. The *Contractor* subcontracts the *Contractor*'s design, the provision of Site services and the construction and installation of the *works* except work which the Contract Data states that he will do himself.

20.3 The *Contractor* advises the *Project Manager* on the practical implications of the design of the *works* and on subcontracting arrangements.

20.4 The *Contractor* prepares forecasts of the total Defined Cost for the whole of the *works* in consultation with the *Project Manager* and submits them to the *Project Manager*. Forecasts are prepared at the intervals stated in the Contract Data from the *starting date* until Completion of the whole of the *works*. An explanation of the changes made since the previous forecast is submitted with each forecast.

 © copyright nec 2005

20.5 If work which the *Contractor* is to do himself is affected by a compensation event, the *Project Manager* and the *Contractor* agree the change to the price for the work and any change to the Completion Date and Key Dates. If they cannot agree, the *Project Manager* decides the change.

Subcontracting 26

26.4 The *Contractor* submits the proposed contract data for each subcontract for acceptance to the *Project Manager* if

- an NEC contract is proposed and
- the *Project Manager* instructs the *Contractor* to make the submission.

A reason for not accepting the proposed contract data is that its use will not allow the *Contractor* to Provide the Works.

Acceleration 36

36.4 When the *Project Manager* accepts a quotation for an acceleration, he changes the Completion Date, the Key Dates and the forecast of the total Defined Cost of the whole of the *works* accordingly and accepts the revised programme.

Assessing the amount due 50

50.7 Payments of Defined Cost made by the *Contractor* in a currency other than the *currency of this contract* are included in the amount due as payments to be made to him in the same currency. Such payments are converted to the *currency of this contract* in order to calculate the Fee using the *exchange rates.*

Defined Cost 52

52.2 The *Contractor* keeps these records

- accounts of payments of Defined Cost,
- proof that the payments have been made,
- communications about and assessments of compensation events for Subcontractors and
- other records as stated in the Works Information.

52.3 The *Contractor* allows the *Project Manager* to inspect at any time within working hours the accounts and records which he is required to keep.

Implementing compensation events 65

65.3 The changes to the forecast amount of the Prices, the Completion Date and the Key Dates are included in the notification implementing a compensation event.

© copyright nec 2005

DISPUTE RESOLUTION

Option W1

Dispute resolution procedure (used unless the United Kingdom Housing Grants, Construction and Regeneration Act 1996 applies).

Dispute resolution **W1**

W1.1 A dispute arising under or in connection with this contract is referred to and decided by the *Adjudicator*.

The *Adjudicator* W1.2 (1) The Parties appoint the *Adjudicator* under the NEC Adjudicator's Contract current at the *starting date*.

(2) The *Adjudicator* acts impartially and decides the dispute as an independent adjudicator and not as an arbitrator.

(3) If the *Adjudicator* is not identified in the Contract Data or if the *Adjudicator* resigns or is unable to act, the Parties choose a new adjudicator jointly. If the Parties have not chosen an adjudicator, either Party may ask the *Adjudicator nominating body* to choose one. The *Adjudicator nominating body* chooses an adjudicator within four days of the request. The chosen adjudicator becomes the *Adjudicator*.

(4) A replacement *Adjudicator* has the power to decide a dispute referred to his predecessor but not decided at the time when the predecessor resigned or became unable to act. He deals with an undecided dispute as if it had been referred to him on the date he was appointed.

(5) The *Adjudicator*, his employees and agents are not liable to the Parties for any action or failure to take action in an adjudication unless the action or failure to take action was in bad faith.

The adjudication W1.3 (1) Disputes are notified and referred to the *Adjudicator* in accordance with the Adjudication Table.

 © copyright nec 2005

ADJUDICATION TABLE

Dispute about	Which Party may refer it to the *Adjudicator*?	When may it be referred to the *Adjudicator*?
An action of the *Project Manager* or the *Supervisor*	The *Contractor*	Between two and four weeks after the *Contractor*'s notification of the dispute to the *Employer* and the *Project Manager*, the notification itself being made not more than four weeks after the *Contractor* becomes aware of the action
The *Project Manager* or *Supervisor* not having taken an action	The *Contractor*	Between two and four weeks after the *Contractor*'s notification of the dispute to the *Employer* and the *Project Manager*, the notification itself being made not more than four weeks after the *Contractor* becomes aware that the action was not taken
A quotation for a compensation event which is treated as having been accepted	The *Employer*	Between two and four weeks after the *Project Manager*'s notification of the dispute to the *Employer* and the *Contractor*, the notification itself being made not more than four weeks after the quotation was treated as accepted
Any other matter	Either Party	Between two and four weeks after notification of the dispute to the other Party and the *Project Manager*

(2) The times for notifying and referring a dispute may be extended by the *Project Manager* if the *Contractor* and the *Project Manager* agree to the extension before the notice or referral is due. The *Project Manager* notifies the extension that has been agreed to the *Contractor*. If a disputed matter is not notified and referred within the times set out in this contract, neither Party may subsequently refer it to the *Adjudicator* or the *tribunal*.

(3) The Party referring the dispute to the *Adjudicator* includes with his referral information to be considered by the *Adjudicator*. Any more information from a Party to be considered by the *Adjudicator* is provided within four weeks of the referral. This period may be extended if the *Adjudicator* and the Parties agree.

(4) If a matter disputed by the *Contractor* under or in connection with a subcontract is also a matter disputed under or in connection with this contract and if the subcontract allows, the *Contractor* may refer the subcontract dispute to the *Adjudicator* at the same time as the main contract referral. The *Adjudicator* then decides the disputes together and references to the Parties for the purposes of the dispute are interpreted as including the Subcontractor.

© copyright nec 2005

(5) The *Adjudicator* may

- review and revise any action or inaction of the *Project Manager* or *Supervisor* related to the dispute and alter a quotation which has been treated as having been accepted,
- take the initiative in ascertaining the facts and the law related to the dispute,
- instruct a Party to provide further information related to the dispute within a stated time and
- instruct a Party to take any other action which he considers necessary to reach his decision and to do so within a stated time.

(6) A communication between a Party and the *Adjudicator* is communicated to the other Party at the same time.

(7) If the *Adjudicator*'s decision includes assessment of additional cost or delay caused to the *Contractor*, he makes his assessment in the same way as a compensation event is assessed.

(8) The *Adjudicator* decides the dispute and notifies the Parties and the *Project Manager* of his decision and his reasons within four weeks of the end of the period for receiving information. This four week period may be extended if the Parties agree.

(9) Unless and until the *Adjudicator* has notified the Parties of his decision, the Parties, the *Project Manager* and the *Supervisor* proceed as if the matter disputed was not disputed.

(10) The *Adjudicator*'s decision is binding on the Parties unless and until revised by the *tribunal* and is enforceable as a matter of contractual obligation between the Parties and not as an arbitral award. The *Adjudicator*'s decision is final and binding if neither Party has notified the other within the times required by this contract that he is dissatisfied with a decision of the *Adjudicator* and intends to refer the matter to the *tribunal*.

(11) The *Adjudicator* may, within two weeks of giving his decision to the Parties, correct any clerical mistake or ambiguity.

Review by the *tribunal* W1.4

(1) A Party does not refer any dispute under or in connection with this contract to the *tribunal* unless it has first been referred to the *Adjudicator* in accordance with this contract.

(2) If, after the *Adjudicator* notifies his decision a Party is dissatisfied, he may notify the other Party that he intends to refer it to the *tribunal*. A Party may not refer a dispute to the *tribunal* unless this notification is given within four weeks of notification of the *Adjudicator*'s decision.

(3) If the *Adjudicator* does not notify his decision within the time provided by this contract, a Party may notify the other Party that he intends to refer the dispute to the *tribunal*. A Party may not refer a dispute to the *tribunal* unless this notification is given within four weeks of the date by which the *Adjudicator* should have notified his decision.

(4) The *tribunal* settles the dispute referred to it. The *tribunal* has the powers to reconsider any decision of the *Adjudicator* and review and revise any action or inaction of the *Project Manager* or the *Supervisor* related to the dispute. A Party is not limited in the *tribunal* proceedings to the information, evidence or arguments put to the *Adjudicator*.

(5) If the *tribunal* is arbitration, the *arbitration procedure*, the place where the arbitration is to be held and the method of choosing the arbitrator are those stated in the Contract Data.

(6) A Party does not call the *Adjudicator* as a witness in *tribunal* proceedings.

 © copyright nec 2005

Option W2

Dispute resolution procedure (used in the United Kingdom when the Housing Grants, Construction and Regeneration Act 1996 applies).

Dispute resolution **W2**

W2.1 (1) A dispute arising under or in connection with this contract is referred to and decided by the *Adjudicator*. A Party may refer a dispute to the *Adjudicator* at any time.

(2) In this Option, time periods stated in days exclude Christmas Day, Good Friday and bank holidays.

The *Adjudicator* W2.2 (1) The Parties appoint the *Adjudicator* under the NEC Adjudicator's Contract current at the *starting date*.

(2) The *Adjudicator* acts impartially and decides the dispute as an independent adjudicator and not as an arbitrator.

(3) If the *Adjudicator* is not identified in the Contract Data or if the *Adjudicator* resigns or becomes unable to act

- the Parties may choose an adjudicator jointly or
- a Party may ask the *Adjudicator nominating body* to choose an adjudicator.

The *Adjudicator nominating body* chooses an adjudicator within four days of the request. The chosen adjudicator becomes the *Adjudicator.*

(4) A replacement *Adjudicator* has the power to decide a dispute referred to his predecessor but not decided at the time when his predecessor resigned or became unable to act. He deals with an undecided dispute as if it had been referred to him on the date he was appointed.

(5) The *Adjudicator,* his employees and agents are not liable to the Parties for any action or failure to take action in an adjudication unless the action or failure to take action was in bad faith.

The adjudication W2.3 (1) Before a Party refers a dispute to the *Adjudicator*, he gives a notice of adjudication to the other Party with a brief description of the dispute and the decision which he wishes the *Adjudicator* to make. If the *Adjudicator* is named in the Contract Data, the Party sends a copy of the notice of adjudication to the *Adjudicator* when it is issued. Within three days of the receipt of the notice of adjudication, the *Adjudicator* notifies the Parties

- that he is able to decide the dispute in accordance with the contract or
- that he is unable to decide the dispute and has resigned.

If the *Adjudicator* does not so notify within three days of the issue of the notice of adjudication, either Party may act as if he has resigned.

(2) Within seven days of a Party giving a notice of adjudication he

- refers the dispute to the *Adjudicator*,
- provides the *Adjudicator* with the information on which he relies, including any supporting documents and
- provides a copy of the information and supporting documents he has provided to the *Adjudicator* to the other Party.

Any further information from a Party to be considered by the *Adjudicator* is provided within fourteen days of the referral. This period may be extended if the *Adjudicator* and the Parties agree.

© copyright nec 2005

(3) If a matter disputed by the *Contractor* under or in connection with a subcontract is also a matter disputed under or in connection with this contract, the *Contractor* may, with the consent of the Subcontractor, refer the subcontract dispute to the *Adjudicator* at the same time as the main contract referral. The *Adjudicator* then decides the disputes together and references to the Parties for the purposes of the dispute are interpreted as including the Subcontractor.

(4) The *Adjudicator* may

- review and revise any action or inaction of the *Project Manager* or *Supervisor* related to the dispute and alter a quotation which has been treated as having been accepted,
- take the initiative in ascertaining the facts and the law related to the dispute,
- instruct a Party to provide further information related to the dispute within a stated time and
- instruct a Party to take any other action which he considers necessary to reach his decision and to do so within a stated time.

(5) If a Party does not comply with any instruction within the time stated by the *Adjudicator*, the *Adjudicator* may continue the adjudication and make his decision based upon the information and evidence he has received.

(6) A communication between a Party and the *Adjudicator* is communicated to the other Party at the same time.

(7) If the *Adjudicator's* decision includes assessment of additional cost or delay caused to the *Contractor*, he makes his assessment in the same way as a compensation event is assessed.

(8) The *Adjudicator* decides the dispute and notifies the Parties and the *Project Manager* of his decision and his reasons within twenty-eight days of the dispute being referred to him. This period may be extended by up to fourteen days with the consent of the referring Party or by any other period agreed by the Parties.

(9) Unless and until the *Adjudicator* has notified the Parties of his decision, the Parties, the *Project Manager* and the *Supervisor* proceed as if the matter disputed was not disputed.

(10) If the *Adjudicator* does not make his decision and notify it to the Parties within the time provided by this contract, the Parties and the *Adjudicator* may agree to extend the period for making his decision. If they do not agree to an extension, either Party may act as if the *Adjudicator* has resigned.

(11) The *Adjudicator*'s decision is binding on the Parties unless and until revised by the *tribunal* and is enforceable as a matter of contractual obligation between the Parties and not as an arbitral award. The *Adjudicator*'s decision is final and binding if neither Party has notified the other within the times required by this contract that he is dissatisfied with a matter decided by the *Adjudicator* and intends to refer the matter to the *tribunal*.

(12) The *Adjudicator* may, within fourteen days of giving his decision to the Parties, correct a clerical mistake or ambiguity.

Review by the *tribunal* W2.4

(1) A Party does not refer any dispute under or in connection with this contract to the *tribunal* unless it has first been decided by the *Adjudicator* in accordance with this contract.

(2) If, after the *Adjudicator* notifies his decision a Party is dissatisfied, that Party may notify the other Party of the matter which he disputes and state that he intends to refer it to the *tribunal*. The dispute may not be referred to the *tribunal* unless this notification is given within four weeks of the notification of the *Adjudicator's* decision.

 © copyright nec 2005

(3) The *tribunal* settles the dispute referred to it. The *tribunal* has the powers to reconsider any decision of the *Adjudicator* and to review and revise any action or inaction of the *Project Manager* or the *Supervisor* related to the dispute. A Party is not limited in *tribunal* proceedings to the information or evidence put to the *Adjudicator*.

(4) If the *tribunal* is arbitration, the *arbitration procedure,* the place where the arbitration is to be held and the method of choosing the arbitrator are those stated in the Contract Data.

(5) A Party does not call the *Adjudicator* as a witness in *tribunal* proceedings.

© copyright nec 2005

SECONDARY OPTION CLAUSES

Option X1: Price adjustment for inflation (used only with Options A, B, C and D)

Defined terms **X1**

X1.1 (a) The Base Date Index (B) is the latest available index before the *base date*.

(b) The Latest Index (L) is the latest available index before the date of assessment of an amount due.

(c) The Price Adjustment Factor is the total of the products of each of the proportions stated in the Contract Data multiplied by (L – B)/B for the index linked to it.

Price Adjustment Factor X1.2 If an index is changed after it has been used in calculating a Price Adjustment Factor, the calculation is repeated and a correction included in the next assessment of the amount due.

The Price Adjustment Factor calculated at the Completion Date for the whole of the *works* is used for calculating price adjustment after this date.

Compensation events X1.3 The Defined Cost for compensation events is assessed using the

- Defined Cost current at the time of assessing the compensation event adjusted to *base date* by dividing by one plus the Price Adjustment Factor for the last assessment of the amount due and
- Defined Cost at *base date* levels for amounts calculated from rates stated in the Contract Data for employees and Equipment.

Price adjustment Options A and B X1.4 Each amount due includes an amount for price adjustment which is the sum of

- the change in the Price for Work Done to Date since the last assessment of the amount due multiplied by the Price Adjustment Factor for the date of the current assessment,
- the amount for price adjustment included in the previous amount due and
- correcting amounts, not included elsewhere, which arise from changes to indices used for assessing previous amounts for price adjustment.

Price adjustment Options C and D X1.5 Each time the amount due is assessed, an amount for price adjustment is added to the total of the Prices which is the sum of

- the change in the Price for Work Done to Date since the last assessment of the amount due multiplied by (PAF/(1 + PAF)) where PAF is the Price Adjustment Factor for the date of the current assessment and
- correcting amounts, not included elsewhere, which arise from changes to indices used for assessing previous amounts for price adjustment.

 © copyright nec 2005

Option X2: Changes in the law

Changes in the law **X2**

X2.1 A change in the law of the country in which the Site is located is a compensation event if it occurs after the Contract Date. The *Project Manager* may notify the *Contractor* of a compensation event for a change in the law and instruct him to submit quotations. If the effect of a compensation event which is a change in the law is to reduce the total Defined Cost, the Prices are reduced.

Option X3: Multiple currencies (used only with Options A and B)

Multiple currencies **X3**

X3.1 The *Contractor* is paid in currencies other than the *currency of this contract* for the items or activities listed in the Contract Data. The *exchange rates* are used to convert from the *currency of this contract* to other currencies.

X3.2 Payments to the *Contractor* in currencies other than the *currency of this contract* do not exceed the maximum amounts stated in the Contract Data. Any excess is paid in the *currency of this contract*.

Option X4: Parent company guarantee

Parent company guarantee **X4**

X4.1 If a parent company owns the *Contractor*, the *Contractor* gives to the *Employer* a guarantee by the parent company of the *Contractor*'s performance in the form set out in the Works Information. If the guarantee was not given by the Contract Date, it is given to the *Employer* within four weeks of the Contract Date.

Option X5: Sectional Completion

Sectional Completion **X5**

X5.1 In these *conditions of contract*, unless stated as the whole of the *works*, each reference and clause relevant to

- the *works*,
- Completion and
- Completion Date

applies, as the case may be, to either the whole of the *works* or any *section* of the *works*.

© copyright nec 2005

Option X6: Bonus for early Completion

Bonus for early Completion **X6**

X6.1 The *Contractor* is paid a bonus calculated at the rate stated in the Contract Data for each day from the earlier of

- Completion and
- the date on which the *Employer* takes over the *works*

until the Completion Date.

Option X7: Delay damages

Delay damages **X7**

X7.1 The *Contractor* pays delay damages at the rate stated in the Contract Data from the Completion Date for each day until the earlier of

- Completion and
- the date on which the *Employer* takes over the *works*.

X7.2 If the Completion Date is changed to a later date after delay damages have been paid, the *Employer* repays the overpayment of damages with interest. Interest is assessed from the date of payment to the date of repayment and the date of repayment is an assessment date.

X7.3 If the *Employer* takes over a part of the *works* before Completion, the delay damages are reduced from the date on which the part is taken over. The *Project Manager* assesses the benefit to the *Employer* of taking over the part of the *works* as a proportion of the benefit to the *Employer* of taking over the whole of the *works* not previously taken over. The delay damages are reduced in this proportion.

Option X12: Partnering

Identified and defined terms **X12**

X12.1 (1) The Partners are those named in the Schedule of Partners. The *Client* is a Partner.

(2) An Own Contract is a contract between two Partners which includes this Option.

(3) The Core Group comprises the Partners listed in the Schedule of Core Group Members.

(4) Partnering Information is information which specifies how the Partners work together and is either in the documents which the Contract Data states it is in or in an instruction given in accordance with this contract.

(5) A Key Performance Indicator is an aspect of performance for which a target is stated in the Schedule of Partners.

 © copyright nec 2005

Actions X12.2 (1) Each Partner works with the other Partners to achieve the *Client's objective* stated in the Contract Data and the objectives of every other Partner stated in the Schedule of Partners.

(2) Each Partner nominates a representative to act for it in dealings with other Partners.

(3) The Core Group acts and takes decisions on behalf of the Partners on those matters stated in the Partnering Information.

(4) The Partners select the members of the Core Group. The Core Group decides how they will work and decides the dates when each member joins and leaves the Core Group. The *Client's* representative leads the Core Group unless stated otherwise in the Partnering Information.

(5) The Core Group keeps the Schedule of Core Group Members and the Schedule of Partners up to date and issues copies of them to the Partners each time either is revised.

(6) This Option does not create a legal partnership between Partners who are not one of the Parties in this contract.

Working together X12.3 (1) The Partners work together as stated in the Partnering Information and in a spirit of mutual trust and co-operation.

(2) A Partner may ask another Partner to provide information which he needs to carry out the work in his Own Contract and the other Partner provides it.

(3) Each Partner gives an early warning to the other Partners when he becomes aware of any matter that could affect the achievement of another Partner's objectives stated in the Schedule of Partners.

(4) The Partners use common information systems as set out in the Partnering Information.

(5) A Partner implements a decision of the Core Group by issuing instructions in accordance with its Own Contracts.

(6) The Core Group may give an instruction to the Partners to change the Partnering Information. Each such change to the Partnering Information is a compensation event which may lead to reduced Prices.

(7) The Core Group prepares and maintains a timetable showing the proposed timing of the contributions of the Partners. The Core Group issues a copy of the timetable to the Partners each time it is revised. The *Contractor* changes his programme if it is necessary to do so in order to comply with the revised timetable. Each such change is a compensation event which may lead to reduced Prices.

(8) A Partner gives advice, information and opinion to the Core Group and to other Partners when asked to do so by the Core Group. This advice, information and opinion relates to work that another Partner is to carry out under its Own Contract and is given fully, openly and objectively. The Partners show contingency and risk allowances in information about costs, prices and timing for future work.

(9) A Partner notifies the Core Group before subcontracting any work.

Incentives X12.4 (1) A Partner is paid the amount stated in the Schedule of Partners if the target stated for a Key Performance Indicator is improved upon or achieved. Payment of the amount is due when the target has been improved upon or achieved and is made as part of the amount due in the Partner's Own Contract.

(2) The *Client* may add a Key Performance Indicator and associated payment to the Schedule of Partners but may not delete or reduce a payment stated in the Schedule of Partners.

© copyright nec 2005

Option X13: Performance bond

Performance bond **X13**

X13.1 The *Contractor* gives the *Employer* a performance bond, provided by a bank or insurer which the *Project Manager* has accepted, for the amount stated in the Contract Data and in the form set out in the Works Information. A reason for not accepting the bank or insurer is that its commercial position is not strong enough to carry the bond. If the bond was not given by the Contract Date, it is given to the *Employer* within four weeks of the Contract Date.

Option X14: Advanced payment to the *Contractor*

Advanced payment **X14**

X14.1 The *Employer* makes an advanced payment to the *Contractor* of the amount stated in the Contract Data.

X14.2 The advanced payment is made either within four weeks of the Contract Date or, if an advanced payment bond is required, within four weeks of the later of

- the Contract Date and
- the date when the *Employer* receives the advanced payment bond.

The advanced payment bond is issued by a bank or insurer which the *Project Manager* has accepted. A reason for not accepting the proposed bank or insurer is that its commercial position is not strong enough to carry the bond. The bond is for the amount of the advanced payment which the *Contractor* has not repaid and is in the form set out in the Works Information. Delay in making the advanced payment is a compensation event.

X14.3 The advanced payment is repaid to the *Employer* by the *Contractor* in instalments of the amount stated in the Contract Data. An instalment is included in each amount due assessed after the period stated in the Contract Data has passed until the advanced payment has been repaid.

Option X15: Limitation of the *Contractor*'s liability for his design to reasonable skill and care

The *Contractor*'s design **X15**

X15.1 The *Contractor* is not liable for Defects in the *works* due to his design so far as he proves that he used reasonable skill and care to ensure that his design complied with the Works Information.

X15.2 If the *Contractor* corrects a Defect for which he is not liable under this contract it is a compensation event.

 © copyright nec 2005

Option X16: Retention (not used with Option F)

Retention **X16**

X16.1 After the Price for Work Done to Date has reached the *retention free amount*, an amount is retained in each amount due. Until the earlier of

- Completion of the whole of the *works* and
- the date on which the *Employer* takes over the whole of the *works*

the amount retained is the *retention percentage* applied to the excess of the Price for Work Done to Date above the *retention free amount*.

X16.2 The amount retained is halved

- in the assessment made at Completion of the whole of the *works* or
- in the next assessment after the *Employer* has taken over the whole of the *works* if this is before Completion of the whole of the *works*.

The amount retained remains at this amount until the Defects Certificate is issued. No amount is retained in the assessments made after the Defects Certificate has been issued.

Option X17: Low performance damages

Low performance damages **X17**

X17.1 If a Defect included in the Defects Certificate shows low performance with respect to a performance level stated in the Contract Data, the *Contractor* pays the amount of low performance damages stated in the Contract Data.

Option X18: Limitation of liability

Limitation of liability **X18**

X18.1 The *Contractor*'s liability to the *Employer* for the *Employer*'s indirect or consequential loss is limited to the amount stated in the Contract Data.

X18.2 For any one event, the liability of the *Contractor* to the *Employer* for loss of or damage to the *Employer*'s property is limited to the amount stated in the Contract Data.

X18.3 The *Contractor*'s liability to the *Employer* for Defects due to his design which are not listed on the Defects Certificate is limited to the amount stated in the Contract Data.

X18.4 The *Contractor*'s total liability to the *Employer* for all matters arising under or in connection with this contract, other than the excluded matters, is limited to the amount stated in the Contract Data and applies in contract, tort or delict and otherwise to the extent allowed under the *law of the contract.*

The excluded matters are amounts payable by the *Contractor* as stated in this contract for

- loss of or damage to the *Employer*'s property,
- delay damages if Option X7 applies,
- low performance damages if Option X17 applies and
- *Contractor*'s share if Option C or Option D applies.

X18.5 The *Contractor* is not liable to the *Employer* for a matter unless it is notified to the *Contractor* before the *end of liability date*.

 © copyright nec 2005

Option X20: Key Performance Indicators (not used with Option X12)

Incentives

X20.1 A Key Performance Indicator is an aspect of performance by the *Contractor* for which a target is stated in the Incentive Schedule. The Incentive Schedule is the *incentive schedule* unless later changed in accordance with this contract.

X20.2 From the *starting date* until the Defects Certificate has been issued, the *Contractor* reports to the *Project Manager* his performance against each of the Key Performance Indicators. Reports are provided at the intervals stated in the Contract Data and include the forecast final measurement against each indicator.

X20.3 If the *Contractor's* forecast final measurement against a Key Performance Indicator will not achieve the target stated in the Incentive Schedule, he submits to the *Project Manager* his proposals for improving performance.

X20.4 The *Contractor* is paid the amount stated in the Incentive Schedule if the target stated for a Key Performance Indicator is improved upon or achieved. Payment of the amount is due when the target has been improved upon or achieved.

X20.5 The *Employer* may add a Key Performance Indicator and associated payment to the Incentive Schedule but may not delete or reduce a payment stated in the Incentive Schedule.

 © copyright nec 2005

OPTION Y

Option Y(UK)2: The Housing Grants, Construction and Regeneration Act 1996

Definitions **Y(UK)2**

Y2.1 (1) The Act is The Housing Grants, Construction and Regeneration Act 1996.

(2) A period of time stated in days is a period calculated in accordance with Section 116 of the Act.

Dates for payment Y2.2 The date on which a payment becomes due is seven days after the assessment date.

The final date for payment is fourteen days or a different period for payment if stated in the Contract Data after the date on which payment becomes due.

The *Project Manager*'s certificate is the notice of payment from the *Employer* to the *Contractor* specifying the amount of the payment made or proposed to be made and stating how the amount was calculated.

Notice of intention to withhold payment Y2.3 If either Party intends to withhold payment of an amount due under this contract, he notifies the other Party not later than seven days (the prescribed period) before the final date for payment by stating the amount proposed to be withheld and the reason for withholding payment. If there is more than one reason, the amount for each reason is stated.

A Party does not withhold payment of an amount due under this contract unless he has notified his intention to withhold payment as required by this contract.

Suspension of performance Y2.4 If the *Contractor* exercises his right under the Act to suspend performance, it is a compensation event.

Option Y(UK)3: The Contracts (Rights of Third Parties) Act 1999

Third party rights **Y(UK)3**

Y3.1 A person or organisation who is not one of the Parties may enforce a term of this contract under the Contracts (Rights of Third Parties) Act 1999 only if the term and the person or organisation are stated in the Contract Data.

Option Z: *Additional conditions of contract*

Additional conditions of contract **Z1**

Z1.1 The *additional conditions of contract* stated in the Contract Data are part of this contract.

© copyright nec 2005

SCHEDULE OF COST COMPONENTS

This schedule is part of the *conditions of contract* only when Option C, D or E is used. In this schedule the *Contractor* means the *Contractor* and not his Subcontractors. An amount is included only in one cost component and only if it is incurred in order to Provide the Works.

People 1 The following components of the cost of

- people who are directly employed by the *Contractor* and whose normal place of working is within the Working Areas and
- people who are directly employed by the *Contractor* and whose normal place of working is not within the Working Areas but who are working in the Working Areas.

11 Wages, salaries and amounts paid by the *Contractor* for people paid according to the time worked while they are within the Working Areas.

12 Payments to people for

(a) bonuses and incentives
(b) overtime
(c) working in special circumstances
(d) special allowances
(e) absence due to sickness and holidays
(f) severance related to work on this contract.

13 Payments made in relation to people for

(a) travel
(b) subsistence and lodging
(c) relocation
(d) medical examinations
(e) passports and visas
(f) travel insurance
(g) items (a) to (f) for dependants
(h) protective clothing
(i) meeting the requirements of the law
(j) pensions and life assurance
(k) death benefit
(l) occupational accident benefits
(m) medical aid
(n) a vehicle
(o) safety training.

14 The following components of the cost of people who are not directly employed by the *Contractor* but are paid for by him according to the time worked while they are within the Working Areas.

Amounts paid by the *Contractor*.

Equipment 2 The following components of the cost of Equipment which is used within the Working Areas (including the cost of accommodation but excluding Equipment cost covered by the percentage for Working Areas overheads).

21 Payments for the hire or rent of Equipment not owned by

- the *Contractor*,
- his parent company or
- by a company with the same parent company

at the hire or rental rate multiplied by the time for which the Equipment is required.

 © copyright nec 2005

22 Payments for Equipment which is not listed in the Contract Data but is

- owned by the *Contractor*,
- purchased by the *Contractor* under a hire purchase or lease agreement or
- hired by the *Contractor* from the *Contractor*'s parent company or from a company with the same parent company

at open market rates, multiplied by the time for which the Equipment is required.

23 Payments for Equipment purchased for work included in this contract listed with a time-related on cost charge, in the Contract Data, of

- the change in value over the period for which the Equipment is required and
- the time-related on cost charge stated in the Contract Data for the period for which the Equipment is required.

The change in value is the difference between the purchase price and either the sale price or the open market sale price at the end of the period for which the Equipment is required. Interim payments of the change in value are made at each assessment date. A final payment is made in the next assessment after the change in value has been determined.

If the *Project Manager* agrees, an additional item of Equipment may be assessed as if it had been listed in the Contract Data.

24 Payments for special Equipment listed in the Contract Data. These amounts are the rates stated in the Contract Data multiplied by the time for which the Equipment is required.

If the *Project Manager* agrees, an additional item of special Equipment may be assessed as if it had been listed in the Contract Data.

25 Payments for the purchase price of Equipment which is consumed.

26 Unless included in the hire or rental rates, payments for

- transporting Equipment to and from the Working Areas other than for repair and maintenance,
- erecting and dismantling Equipment and
- constructing, fabricating or modifying Equipment as a result of a compensation event.

27 Payments for purchase of materials used to construct or fabricate Equipment.

28 Unless included in the hire rates, the cost of operatives is included in the cost of people.

Plant and Materials **3** The following components of the cost of Plant and Materials.

31 Payments for

- purchasing Plant and Materials,
- delivery to and removal from the Working Areas,
- providing and removing packaging and
- samples and tests.

32 Cost is credited with payments received for disposal of Plant and Materials unless the cost is disallowed.

Charges **4** The following components of the cost of charges paid by the *Contractor.*

41 Payments for provision and use in the Working Areas of

- water,
- gas and
- electricity.

© copyright nec 2005

42 Payments to public authorities and other properly constituted authorities of charges which they are authorised to make in respect of the *works*.

43 Payments for

(a) cancellation charges arising from a compensation event
(b) buying or leasing land
(c) compensation for loss of crops or buildings
(d) royalties
(e) inspection certificates
(f) charges for access to the Working Areas
(g) facilities for visits to the Working Areas by Others
(h) specialist services
(i) consumables and equipment provided by the *Contractor* for the *Project Manager*'s and *Supervisor*'s offices.

44 A charge for overhead costs incurred within the Working Areas calculated by applying the percentage for Working Areas overheads stated in the Contract Data to the total of people items 11, 12, 13 and 14. The charge includes provision and use of equipment, supplies and services, but excludes accommodation, for

(a) catering
(b) medical facilities and first aid
(c) recreation
(d) sanitation
(e) security
(f) copying
(g) telephone, telex, fax, radio and CCTV
(h) surveying and setting out
(i) computing
(j) hand tools not powered by compressed air.

Manufacture and fabrication

5 The following components of the cost of manufacture and fabrication of Plant and Materials which are

- wholly or partly designed specifically for the *works* and
- manufactured or fabricated outside the Working Areas.

51 The total of the hours worked by employees multiplied by the hourly rates stated in the Contract Data for the categories of employees listed.

52 An amount for overheads calculated by multiplying this total by the percentage for manufacturing and fabrication overheads stated in the Contract Data.

Design

6 The following components of the cost of design of the *works* and Equipment done outside the Working Areas.

61 The total of the hours worked by employees multiplied by the hourly rates stated in the Contract Data for the categories of employees listed.

62 An amount for overheads calculated by multiplying this total by the percentage for design overheads stated in the Contract Data.

63 The cost of travel to and from the Working Areas for the categories of design employees listed in the Contract Data.

Insurance

7 The following are deducted from cost

- the cost of events for which this contract requires the *Contractor* to insure and
- other costs paid to the *Contractor* by insurers.

 © copyright nec 2005

SHORTER SCHEDULE OF COST COMPONENTS

This schedule is part of the *conditions of contract* only when Option A, B, C, D or E is used. When Option C, D or E is used, this schedule is used by agreement for assessing compensation events. When Option C, D or E is used, in this schedule the *Contractor* means the *Contractor* and not his Subcontractors. An amount is included only in one cost component and only if it is incurred in order to Provide the Works.

People **1** The following components of the cost of

- people who are directly employed by the *Contractor* and whose normal place of working is within the Working Areas,
- people who are directly employed by the *Contractor* and whose normal place of working is not within the Working Areas but who are working in the Working Areas and
- people who are not directly employed by the *Contractor* but are paid for by him according to the time worked while they are within the Working Areas.

11 Amounts paid by the *Contractor* including those for meeting the requirements of the law and for pension provision.

Equipment **2** The following components of the cost of Equipment which is used within the Working Areas (including the cost of accommodation but excluding Equipment cost covered by the percentage for people overheads).

21 Amounts for Equipment which is in the published list stated in the Contract Data. These amounts are calculated by applying the percentage adjustment for listed Equipment stated in the Contract Data to the rates in the published list and by multiplying the resulting rate by the time for which the Equipment is required.

22 Amounts for Equipment listed in the Contract Data which is not in the published list stated in the Contract Data. These amounts are the rates stated in the Contract Data multiplied by the time for which the Equipment is required.

23 The time required is expressed in hours, days, weeks or months consistently with the list of items of Equipment in the Contract Data or with the published list stated in the Contract Data.

24 Unless the item is in the published list and the rate includes the cost component, payments for

- transporting Equipment to and from the Working Areas other than for repair and maintenance,
- erecting and dismantling Equipment and
- constructing, fabricating or modifying Equipment as a result of a compensation event.

25 Unless the item is in the published list and the rate includes the cost component, the purchase price of Equipment which is consumed.

26 Unless included in the rate in the published list, the cost of operatives is included in the cost of people.

27 Amounts for Equipment which is neither in the published list stated in the Contract Data nor listed in the Contract Data, at competitively tendered or open market rates, multiplied by the time for which the Equipment is required.

© copyright nec 2005

Plant and Materials **3** The following components of the cost of Plant and Materials.

31 Payments for

- purchasing Plant and Materials,
- delivery to and removal from the Working Areas,
- providing and removing packaging and
- samples and tests.

32 Cost is credited with payments received for disposal of Plant and Materials unless the cost is disallowed.

Charges **4** The following components of the cost of charges paid by the *Contractor.*

41 A charge calculated by applying the percentage for people overheads stated in the Contract Data to people item 11 to cover the costs of

- payments for the provision and use in the Working Areas of water, gas and electricity,
- payments for buying or leasing land, compensation for loss of crops or buildings, royalties, inspection certificates, charges for access to the Working Areas, facilities for visits to the Working Areas by Others and
- payments for equipment, supplies and services for offices, drawing office, laboratories, workshops, stores and compounds, labour camps, cabins, catering, medical facilities and first aid, recreation, sanitation, security, copying, telephone, telex, fax, radio, CCTV, surveying and setting out, computing, and hand tools not powered by compressed air.

42 Payments for cancellation charges arising from a compensation event.

43 Payments to public authorities and other properly constituted authorities of charges which they are authorised to make in respect of the *works.*

44 Consumables and equipment provided by the *Contractor* for the *Project Manager*'s and *Supervisor*'s office.

45 Specialist services.

Manufacture and fabrication **5** The following components of the cost of manufacture and fabrication of Plant and Materials, which are

- wholly or partly designed specifically for the *works* and
- manufactured or fabricated outside the Working Areas.

51 Amounts paid by the *Contractor.*

Design **6** The following components of the cost of design of the *works* and Equipment done outside the Working Areas.

61 The total of the hours worked by employees multiplied by the hourly rates stated in the Contract Data for the categories of employees listed.

62 An amount for overheads calculated by multiplying this total by the percentage for design overheads stated in the Contract Data.

63 The cost of travel to and from the Working Areas for the categories of design employees listed in the Contract Data.

Insurance **7** The following are deducted from cost

- costs against which this contract required the *Contractor* to insure and
- other costs paid to the *Contractor* by insurers.

 © copyright nec 2005

CONTRACT DATA

Part one – Data provided by the *Employer*

Completion of the data in full, according to the Options chosen, is essential to create a complete contract.

Statements given in all contracts

1 General

- The *conditions of contract* are the core clauses and the clauses for main Option . . . . . , dispute resolution Option . . . . . and secondary Options . . . . . of the NEC3 Engineering and Construction Contract June 2005 (with amendments June 2006).
- The *works* are

  . . . . . .
- The *Employer* is

  Name . . . . . .

  Address . . . . . .

  . . . . . .
- The *Project Manager* is

  Name . . . . . .

  Address . . . . . .

  . . . . . .
- The *Supervisor* is

  Name . . . . . .

  Address . . . . . .

  . . . . . .
- The *Adjudicator* is

  Name . . . . . .

  Address . . . . . .

  . . . . . .
- The Works Information is in

  . . . . . .

  . . . . . .

  . . . . . .
- The Site Information is in

  . . . . . .

  . . . . . .

  . . . . . .

© copyright nec 2005

- The *boundaries of the site* are. . . . . .
- The *language of this contract* is . . . . . .
- The *law of the contract* is the law of . . . . . .
- The *period for reply* is . . . . . weeks.
- The *Adjudicator nominating body* is . . . . . .
- The *tribunal* is . . . . . .

  . . . . . .
- The following matters will be included in the Risk Register

  . . . . . .

  . . . . . .

  . . . . . .

3 Time

- The *starting date* is . . . . . .
- The *access dates* are

Part of the Site	Date
1	. . . . .
2	. . . . .
3	. . . . .

- The *Contractor* submits revised programmes at intervals no longer than . . . . . weeks.

4 Testing and Defects

- The *defects date* is . . . . . weeks after Completion of the whole of the *works*.
- The *defect correction period* is . . . . . weeks except that
  - The *defect correction period* for . . . . . is . . . . . weeks
  - The *defect correction period* for . . . . . is . . . . . weeks.

5 Payment

- The *currency of this contract* is the . . . . . .
- The *assessment interval* is . . . . . weeks (not more than five).
- The *interest rate* is . . . . . % per annum (not less than 2) above the . . . . . rate of the . . . . . bank.

 © copyright nec 2005

6 Compensation events

- The place where weather is to be recorded is

  . . . . . .

- The *weather measurements* to be recorded for each calendar month are
  - the cumulative rainfall (mm)
  - the number of days with rainfall more than 5 mm
  - the number of days with minimum air temperature less than 0 degrees Celsius
  - the number of days with snow lying at . . . . . hours GMT
  - and these measurements:

    . . . . . .

    . . . . . .

    . . . . . .

- The *weather measurements* are supplied by . . . . . .
- The *weather data* are the records of past *weather measurements* for each calendar month which were recorded at . . . . . and which are available from . . . . . .

  . . . . . .

Where no recorded data are available

- Assumed values for the ten year return *weather data* for each *weather measurement* for each calendar month are

  . . . . . .

  . . . . . .

  . . . . . .

  . . . . . .

8 Risks and insurance

- The minimum limit of indemnity for insurance in respect of loss of or damage to property (except the *works*, Plant and Materials and Equipment) and liability for bodily injury to or death of a person (not an employee of the *Contractor*) caused by activity in connection with this contract for any one event is

  . . . . . .

- The minimum limit of indemnity for insurance in respect of death of or bodily injury to employees of the *Contractor* arising out of and in the course of their employment in connection with this contract for any one event is

  . . . . . .

Optional statements

If the *tribunal* is arbitration

- The *arbitration procedure* is . . . . . .
- The place where arbitration is to be held is

  . . . . . .

- The person or organisation who will choose an arbitrator
  - if the Parties cannot agree a choice or
  - if the *arbitration procedure* does not state who selects an arbitrator is

  . . . . . .

© copyright nec 2005

If the *Employer* has decided the *completion date* for the whole of the *works*

- The *completion date* for the whole of the *works* is . . . . . .

If the *Employer* is not willing to take over the *works* before the Completion Date

- The *Employer* is not willing to take over the *works* before the Completion Date.

If no programme is identified in part two of the Contract Data

- The *Contractor* is to submit a first programme for acceptance within . . . . . weeks of the Contract Date.

If the *Employer* has identified work which is to meet a stated *condition* by a *key date*

- The *key dates* and *conditions* to be met are

condition to be met	*key date*
1	. . . . .
2	. . . . .
3	. . . . .

If the period in which payments are made is not three weeks and Y(UK)2 is not used

- The period within which payments are made is . . . . . .

If Y(UK)2 is used and the final date for payment is not 14 days after the date when payment is due

- The period for payment is . . . . . .

If there are additional *Employer*'s risks

- These are additional *Employer*'s risks

  1 . . . . . .

  2 . . . . . .

  3 . . . . . .

If the *Employer* is to provide Plant and Materials

- The insurance against loss of or damage to the *works*, Plant and Materials is to include cover for Plant and Materials provided by the *Employer* for an amount of

  . . . . . .

 © copyright nec 2005

If the *Employer* is to provide any of the insurances stated in the Insurance Table

- The *Employer* provides these insurances from the Insurance Table

  1. Insurance against . . . . . .

  Cover/indemnity is . . . . . .

  The deductibles are. . . . . .

  2. Insurance against . . . . . .

  Cover/indemnity is . . . . . .

  The deductibles are. . . . . .

  3. Insurance against . . . . . .

  Cover/indemnity is . . . . . .

  The deductibles are. . . . . .

If additional insurances are to be provided

- The *Employer* provides these additional insurances

  1. Insurance against . . . . . .

  Cover/indemnity is . . . . . .

  The deductibles are. . . . . .

  2. Insurance against . . . . . .

  Cover/indemnity is . . . . . .

  The deductibles are. . . . . .

  3. Insurance against . . . . . .

  Cover/indemnity is . . . . . .

  The deductibles are. . . . . .

- The *Contractor* provides these additional insurances

  1. Insurance against . . . . . .

  Cover/indemnity is . . . . . .

  2. Insurance against . . . . . .

  Cover/indemnity is . . . . . .

  3. Insurance against . . . . . .

  Cover/indemnity is . . . . . .

If Option B or D is used

- The *method of measurement* is . . . . . .

  amended as follows . . . . . .

  . . . . . .

  . . . . . .

© copyright nec 2005

If Option C or D is used

- The *Contractor's share percentages* and the *share ranges* are

share range	*Contractor's share percentage*
less than ..... %	..... %
from .....% to ..... %	..... %
from .....% to ..... %	..... %
greater than..... %	.....%.

If Option C, D, E or F is used

- The *Contractor* prepares forecasts of Defined Cost for the *works* at intervals no longer than..... weeks.
- The *exchange rates* are those published in.....
  on ..... (date).

If Option X1 is used

- The proportions used to calculate the Price Adjustment Factor are

  0. ..... linked to the index for .....
  0. ....
  0. ....
  0. ....
  0. ....
  0. ....
  0. ..... non-adjustable

  1.00

- The *base date* for indices is .....
- The indices are those prepared by.....

If Option X3 is used

- The *Employer* will pay for the items or activities listed below in the currencies stated

items and activities	other currency	total maximum payment in the currency
.....	.....	.....
.....	.....	.....
.....	.....	.....
.....	.....	.....

- The *exchange rates* are those published in .....
  on ..... (date).

 © copyright nec 2005

If Option X5 is used

- The *completion date* for each *section* of the *works* is

section	description	*completion date*
1	.....	.....
2	.....	.....
3	.....	.....
4	.....	.....

If Options X5 and X6 are used together

- The bonus for each *section* of the *works* is

section	description	amount per day
1	.....	.....
2	.....	.....
3	.....	.....
4	.....	.....

Remainder of the *works* .....

If Options X5 and X7 are used together

- Delay damages for each *section* of the *works* are

section	description	amount per day
1	.....	.....
2	.....	.....
3	.....	.....
4	.....	.....

Remainder of the *works* .....

If Option X6 is used (but not if Option X5 is also used)

- The bonus for the whole of the *works* is ..... per day.

If Option X7 is used (but not if Option X5 is also used)

- Delay damages for Completion of the whole of the *works* are ..... per day.

If Option X12 is used

- The *Client* is

Name .....

Address .....

.....

- The *Client's objective* is

.....

.....

.....

.....

.....

.....

© copyright nec 2005

- The Partnering Information is in

  .....

  .....

  .....

  .....

If Option X13 is used

- The amount of the performance bond is .....

If Option X14 is used

- The amount of the advanced payment is .....
- The *Contractor* repays the instalments in assessments starting not less than ..... weeks after the Contract Date.
- The instalments are.....

  .....

  (either an amount or a percentage of the payment otherwise due)
- An advanced payment bond is/is not required.

If Option X16 is used

- The *retention free amount* is.....
- The *retention percentage* is .....%.

If Option X17 is used

- The amounts for low performance damages are

amount	performance level
.....	for .....
.....	for .....
.....	for .....
.....	for .....

If Option X18 is used

- The *Contractor*'s liability to the *Employer* for indirect or consequential loss is limited to .....
- For any one event, the *Contractor*'s liability to the *Employer* for loss of or damage to the *Employer*'s property is limited to.....
- The *Contractor*'s liability for Defects due to his design which are not listed on the Defects Certificate is limited to .....
- The *Contractor*'s total liability to the *Employer* for all matters arising under or in connection with this contract, other than excluded matters, is limited to.....
- The *end of liability date* is.....years after the Completion of the whole of the *works*.

If Option X20 is used (but not if Option X12 is also used)

- The *incentive schedule* for Key Performance Indicators is in .....
- A report of performance against each Key Performance Indicator is provided at intervals of ..... months.

 © copyright nec 2005

If Option Y(UK)3 is used

- term

  .....

  .....

  .....

  .....

- person or organisation

  .....

  .....

  .....

  .....

If Option Z is used

- The *additional conditions of contract* are .....

  .....

© copyright nec 2005

Part two – Data provided by the *Contractor*

Completion of the data in full, according to the Options chosen, is essential to create a complete contract.

Statements given in all contracts

- The *Contractor* is

  Name .....

  Address .....

  .....
- The *direct fee percentage* is ..... %.
- The *subcontracted fee percentage* is ..... %.
- The *working areas* are the Site and .....
- The key people are

  (1) Name.....

  Job.....

  Responsibilities.....

  .....

  Qualifications .....

  Experience.....

  .....

  (2) Name.....

  Job.....

  Responsibilities.....

  .....

  Qualifications.....

  Experience.....

  .....
- The following matters will be included in the Risk Register

  .....

  .....

  .....

  .....

Optional statements

If the *Contractor* is to provide Works Information for his design

- The Works Information for the *Contractor*'s design is in

  .....

  .....

  .....

  .....

  .....

  .....

 © copyright nec 2005

If a programme is to be identified in the Contract Data

- The programme identified in the Contract Data is .....

If the *Contractor* is to decide the *completion date* for the whole of the *works*

- The *completion date* for the whole of the *works* is .....

If Option A or C is used

- The *activity schedule* is .....

If Option B or D is used

- The *bill of quantities* is.....

If Option A, B, C or D is used

- The tendered total of the Prices is .....

If Option F is used

- Work which the *Contractor* will do himself is

activity	*price* (lump sum or unit rate)
.....	.....
.....	.....
.....	.....
.....	.....

If Option A or B is used

Data for the Shorter Schedule of Cost Components

- The percentage for people overheads is.....%.
- The published list of Equipment is the last edition of the list published by .....
- The percentage for adjustment for Equipment in the published list is ..... % (state plus or minus).
- The rates for other Equipment are

Equipment	size or capacity	rate
.....	.....	.....
.....	.....	.....
.....	.....	.....
.....	.....	.....

- The hourly rates for Defined Cost of design outside the Working Areas are

category of employee	hourly rate
.....	.....
.....	.....
.....	.....
.....	.....

- The percentage for design overheads is.....%.

© copyright nec 2005

- The categories of design employees whose travelling expenses to and from the Working Areas are included in Defined Cost are

  .....

  .....

  .....

  .....

If Option C, D or E is used

Data for Schedule of Cost Components

- The listed items of Equipment purchased for work on this contract, with an on cost charge, are

Equipment	time-related charge	per time period
.....	.....	per .....
.....	.....	per .....
.....	.....	per .....
.....	.....	per .....

- The rates for special Equipment are

Equipment	size or capacity	rate
.....	.....	.....
.....	.....	.....
.....	.....	.....
.....	.....	.....

- The percentage for Working Areas overheads is .....%.
- The hourly rates for Defined Cost of manufacture and fabrication outside the Working Areas are

category of employee	hourly rate
.....	.....
.....	.....
.....	.....
.....	.....

- The percentage for manufacture and fabrication overheads is .....%.

If Option C, D or E is used

Data for both schedules of cost components

- The hourly rates for Defined Cost of design outside the Working Areas are

category of employee	hourly rate
.....	.....
.....	.....
.....	.....
.....	.....

- The percentage for design overheads is.....%.

 © copyright nec 2005

- The categories of design employees whose travelling expenses to and from the Working Areas are included as a cost of design of the *works* and Equipment done outside of the Working Areas are

  .....

  .....

  .....

  .....

If Option C, D or E is used

Data for the Shorter Schedule of Cost Components

- The percentage for people overheads is..... %.
- The published list of Equipment is the last edition of the list published by

  .....
- The percentage for adjustment for Equipment in the published list is

  ..... % (state plus or minus).
- The rates for other Equipment are

Equipment	size or capacity	rate
.....	.....	.....
.....	.....	.....
.....	.....	.....
.....	.....	.....

© copyright nec 2005

Index by clause numbers (Option clauses indicated by their letters, main clause heads by bold numbers). Terms in *italics* are identified in Contract Data, and defined terms have capital initial letters.

acceleration **36**, A36.3, B36.3, C36.3, D36.3, E36.4, F36.4
  compensation events 60.1(9)
  Key Date 36.1
  quotations 36.1, B11.2(21), D11.2(21)
acceptance
  addition to Working Areas 15.1
  conditions of subcontract 26.3
  contract data for subcontract C26.4, D26.4, E26.4, F26.4
  Contractor's communication 13.4, 13.8, 14.1
  Contractor's design 21.2, 21.3
  Defects **44**, 60.1(1)(9)
  Equipment design 23.1
  form of communication 13.1
  insurance policies/certificates 85.1, 87.1, 87.2
  programme (first/revised) 31.3, 32.2, 50.3, 64.1–2, A31.4, A36.3, B36.3, C31.4, C36.3, D36.3
  quotations 44.2, 60.1(9), 62.3, 65.1, A36.3, A65.4, B36.3, C36.3, D36.3, E36.4, F36.4
  reasons for withholding 13.4, 13.8, 15.1, 21.2, 23.1, 24.1, 26.2–3, 31.3, 60.1(9), 85.1, 87.1, A54.3, C26.4, D26.4, E26.4, F26.4, X13.1
  replacement *Contractor*'s employee 24.1
  subcontract conditions 26.3
  Subcontractor 26.2, 91.2
acceptance procedure C11.2(25), D11.2(25), E11.2(25)
  Works Information F11.2(26)
Accepted Programme 62.2
  absence 64.2
  and Activity Schedule A54.2, A54.3, C54.2, C54.3
  changes 32.1, 63.7
  dates/times 33.1, 60.1(2)(3)(5), 63.3
  definition 11.2(1)
  prevention 19.1, 60.1(19)
  termination 91.7
  see also programme
access 27.2, 43.4
  to Site 31.2
  and use of Site **33**
access date 30.1, 31.2
  Site 33.1
accounts and records, *Contractor*'s C11.2(25), C52.2, C52.3, D11.2(25), D52.2–3, E11.2(25), E52.2, E52.3, F11.2(26), F52.2, F52.3
actions **10**, 11.2(13,14), 14.2, 16.3, X12.2
activity schedule A11.2(20), C11.2(20)
Activity Schedule A11.2(20,30), A31.4, A54.1–3, A63.12, C11.2(20)(30), C31.4, C54.1–3, C63.12
  revised A54.2, A54.3, C54.2, C54.3
Acts
  Contracts (Rights of Third Parties) (1999) **Y(UK)3**
  Housing Grants, Construction and Regeneration (1996) **W1**, **W2**, **Y(UK)2**
additional conditions of contract **Z1**
adjudication
  dispute resolution W1.3(1–11), W2.3(1–12)
  timings W1.3(8), W1.3(11), W2.3(2)(8)(11–12)
Adjudication Table W1.3(1)
Adjudicator
  decisions 51.3
  dispute resolution W1.2(1-5), W2.2(1–5)
  nominating body W1.2(3), W2.2(3)
  not arbitrator W1.2(2), W2.2(2)
  replacement W1.2(4), W2.2(4)
  resignation of W1.2(3)
  tribunal W1.3(2)(10), W2.3(11)
  tribunal reviews W1.4(1–4)(6), W2.4(1–3)
administration order 91.1
advanced payment 93.1, X14.1–3
  repayment X14.3
advanced payment bond X14.2
advanced payment(s) 93.1, X14.1–3
agents of *Adjudicator* W1.2(5), W2.2(5)
agreement, Parties/Party 12.4
ambiguities **17**, 60.3, 63.8, B60.6, D60.6
amount due
  adjustment for inflation X1.1(b), X1.2, X1.4, X1.5
  and advanced payment repayment 93.1, X14.3
  assessing **50**, 51.1, 51.3, 90.4, 93.1, C50.6, C53.3–4, D50.6, E50.7, F50.7, X1.1(b), X1.2, X1.4
  Contractor's share D53.7–8
  correction of assessment 50.5, 51.3, X1.2
  definition 50.2
  payment 51.1, 51.3
  retention in 93.1, X16.1
  on termination 90.2, 90.4, 93.1, 93.2, A93.3
amount retained 50.3, X16.1, X16.2
applicable law 11.2(5), 21.2, 23.1, 84.2
  tests and inspections 40.1
approval from Others, *Contractor*'s design 27.1
arbitration procedure W1.4(5), W2.4(4)
arrangement with creditors 91.1
assessment
  Adjudicator W1.3(7), W2.3(7)
  amount due **50**, 51.1, 51.3, 90.4, 93.1, C50.6, C53.3, C53.4, D50.6, D53.7, D53.8, E50.7, F50.7, X1.1(b), X1.2, X1.4, X1.5
  compensation events 62.2–3, **63**, **64**, 65.1–2, A63.10, A63.12, A65.4, B63.10–13, C52.2, C63.11–12, C63.15, D11.2(25), D52.2, D63.13, D63.15, E11.2(25), E52.2, E63.15, F52.2, X1.3

 © copyright nec 2005

Contractor's share C53.1, C53.3–4, C93.4, C93.6, D53.5, D53.7–8, D93.5–6
Defect(s) correction cost 45.2
interest 51.2, 51.3
tests/inspections cost 40.6
assessment date 50.1, 50.4, 51.1–2, C11.2(29), D11.2(29), E11.2(29), F11.2(29), X7.2, Y2.2
assessment interval 50.1
assignment of benefit 92.2
availability of resources C11.2(25), D11.2(25), E11.2(25)

bank bonds X13.1, X14.2
bankruptcy 91.1
base date X1.1(a), X1.3
Base Date Index X1.1(a)
bill of quantities B11.2(21), D11.2(21)
Bill of Quantities B11.2(21)(28)(31), B55.1, B60.4–7, B63.13, D11.2(21)(31)(33), D55.1, D60.4–7, D63.13
binding decisions, *Adjudicator* W1.3(10), W2.3(11)
bodily injury 84.2
bond(s) 91.2, X13.1, X14.2
bonus for early Completion **X6**
boundaries of the site 11.2(15)
breach of contract 60.1(18)
breach of statutory duty 80.1

calculations, assessment of compensation events for Subcontractors C52.2, D52.2, E52.2, F52.2
certificate(s)/certification
communication of 13.1, 13.6
Completion 30.2, 35.2
Defects 43.3, 50.1
payment 50.5, 51.1–3, 90.4, 91.4
take-over date 35.3, 60.1(15)
termination 90.1, 90.3, 90.5, C93.4, D93.5
changed Completion Date X7.2
changed decision(s) 60.1(8), 61.1–3
change(s) in Activity Schedule A54.2, A63.12, C54.2, C63.12
change(s) in forecast Prices, Completion Date and Key Dates due to compensation events E65.3, F65.3
change(s) in the law **X2**
change(s) of Prices, Completion Date and Key Dates due to compensation events A65.4, B65.4, C65.4, D65.4
change(s) of Prices due to compensation events 62.2, 63.1, A63.12, B60.6, B63.13, C63.11, D60.6, D63.13
change(s) in Works Information 14.3, 18.1, 44.1, 44.2, 60.1(1), 63.8–9
civil commotion/war risks 80.1
claims
Employer's risk 80.1
indemnity against 83.1, 83.2
co-operation at risk reduction meeting 16.3
co-operation with Others 25.1–3
co-operation and trust 10.1, 26.3, X12.3(1)
communications **13**
and acceptance of submission 13.4, 13.8, 14.1
address to which sent 13.2
Adjudicator W1.3(6), W2.3(6)
assessment of compensation events for Subcontractors C52.2, D52.2, E52.2, F52.2
certificate(s) 13.6
form defined 13.1
language for 13.1
notification(s) 13.7
period for reply 13.3, 13.5
receipt of 13.2
time allowed for reply 13.3, 13.5, 31.3, 32.2, 60.1(6)
time when effective 13.2
compensation
Employer's risk 80.1
see also indemnity
compensation events **60**
amount due affected by 51.3
assessing 62.2–3, **63**, **64**, 65.1–2, A63.10, A63.12, A65.4, B63.10–13, C52.2, C63.11–12, C63.15, D11.2(25), D52.2, D63.13, D63.15, E11.2(25), E52.2, E63.15, F52.2, X1.3
assumptions made 60.1(17), 61.6, 63.7, 65.2
Completion Date affected by 62.2, A65.4, B65.4, C65.4, D65.4
Completion Date notification 61.3
Contractor 60.1(1)(8)(10)(12)(19), 60.2, 60.3, 61.1–6, F20.5
and Defined Cost 61.4
definitions 13.8, 60.1, B60.4–7, D60.4–D60.7, X2.1, X14.2
implementing 32.1, **65**, A65.4, B65.4, C65.4, D65.4, E65.3, F65.3
Key Dates affected by 60.1(4), 61.3, 62.2, A65.4, B60.5, B65.4, C65.4, D60.5, D65.4
notifying **61**, X2.1
payment to Subcontractor D11.2(25), C11.2(25), E11.2(25), F11.2(26)
performance suspension Y2.4
Prices affected by 62.2, 63.1, A63.10, A65.4, B60.6, B63.10, B65.4, C65.4, D60.4, D60.6, D63.13, D65.4, X2.1
Prices not affected by 61.4
Project Manager's assessments **64**, A63.10, A65.4, B63.10, C63.15, D63.15, E63.15
quotations 61.4, 61.5, **62**, 64.1, 64.3–4, 65.1, A11.2(22), B11.2(22), C11.2(25), D11.2(25), E11.2(25), F11.2(26), X2.1
completed activities A11.2(27)
completed work, definition B11.2(28)
Completion **30**
acceleration of 36.1
bonus paid if early X6.1
certification of 30.2, 35.2
and compensation events 61.4, B60.5, D60.5
decision on date 30.2
definition 11.2(2)
early warning of possible delay 16.1
Key Date 16.1
and programme 31.2, 32.2
section of the *works* X5.1

 © copyright nec 2005

Completion (*continued*)
and take over of *works* 35.1–2, 35.3, 60.1(15), X16.1, X16.2
whole of *works* 32.2, 50.1, 60.1(13), 92.1, 93.2, C20.4, C53.3, D20.4, D53.7, E20.4, F20.4, X1.2, X16.1
work necessary 11.2(2)(13)
Completion Date (*completion date*)
accelerated programme 36.1, A36.3, B36.3, C36.3, D36.3, E36.4, F36.4
change to later date X7.2
and compensation events 60.1(13)(15), 61.4, 62.2, 63.3, A65.4, B65.4, C65.4, D65.4, E65.3, F65.3
definition 11.2(3)
if Defect(s) accepted 44.2
notification of compensation events 61.3
Price Adjustment Factor calculation X1.2
in programme (first/revised) 31.2, 36.1, A36.3, B36.3, C36.3, D36.3
rights to change 63.4
section of the *works* X5.1
and take over of *works* 35.1–2, 60.1(15)
and *weather measurement* 60.1(13)
work to be done by 11.2(2), 30.1
Condition 11.2(9), 25.3, 30.3, 31.2, 63.3, 63.9, B60.5, D60.5
conditions, identified and defined terms 11.2(9)
conditions of contract
identified and defined terms 11.1
for Subcontract 26.3
conditions of contract
additional Z1.1
interpretation 12.3
and sectional Completion X5.1
Contract Data/contract data
addresses for communications 13.2
Adjudicator identified W1.2(3), W2.2(3)
conditions of contract Z1.1
conditions of contract 11.1
currencies for payment X3.1, X3.2
delay damages X7.1
documents referred to in 11.2(16)(19)
early Completion bonus X6.1
Employer's risks 80.1
forecasts of Defined Cost C20.4, D20.4, E20.4, F20.4
insurance 84.1, 87.2
low performance damages X17.1
named key person 24.1
payment 51.2, X14.1, X14.3
performance bond X13.1
performance level X17.1
price adjustment for inflation X1.1(c), X1.3
programme 11.2(1), 31.1, 50.3
start-to-Completion interval 32.2
for subcontract C26.4, D26.4, E26.4, F26.4
take over of *works* 35.1
terms identified 11.1
third party rights Y3.1
time allowances 13.4, 31.1
weather data/weather measurement 60.1(13)
Contract Date 11.2(4), B60.4, D60.4, X2.1, X4.1, X13.1, X14.2
Accepted Programme 60.1(19)
compensation events 60.1(12)(19)
prevention 19.1
termination 91.7
Contractor
acceleration 36.1, 36.2
acceptance of communication/proposal/submission 13.4, 13.8, 14.1, 15.1, 21.2, 87.2
access 27.2, 43.4
access and use of Site 33.1
accounts and records C11.2(25), C52.2–3, D11.2(25), D52.2–3, E11.2(25), E52.2, E52.3, F11.2(26), F52.2, F52.3
actions 10.1
Activity Schedule A31.4, A54.2, C31.4, C54.2
adding to *working areas* 15.1
adjudication W1.3(1–2)(4), W2.3(3)(7)
advanced payment X14.1, X14.3
advice to *Project Manager* C20.3, D20.3, E20.3, F20.3
and ambiguities/inconsistencies 17.1
amount due to be paid to 50.2, 50.4, 51.1
assessment of compensation events 62.2, 64.1, B63.13, C63.15, D63.13, D63.15, E63.15
Bill of Quantities B11.2(28), B63.13, D63.13
certificates issued to 13.6
co-operation 10.1, 25.1–3
compensation events 60.1(1)(8)(10)(12)(19), 60.2, 60.3, 61.1–6, F20.5
costs 52.1, C40.7, D40.7
decisions communicated to 30.2, 50.1, 60.1(8)
default(s) 91.2, 91.3, 91.6
Defect(s) 11.2(5), 42.2
Defect(s) acceptance/non-correction 44.1, 44.2
Defect(s) correction 11.2(2)(6), 43.1–2, 43.4, 45.1–2, C11.2(25), D11.2(25), E11.2(25)
delay damages X7.1
delivery of Plant and Materials 41.1, 93.1
early Completion bonus X6.1
early warning 16.1, 16.2, 16.4, 61.5, 63.5
Equipment 11.2(7), 23.1, 31.2, 92.2
facilities and services provided/paid for by 40.2
fault 61.1, 61.4, C11.2(25), D11.2(25), E11.2(25), F11.2(26)
health and safety 27.4
illegal/impossible requirements 18.1
implementing compensation events 32.1, 65.1, E65.3, F65.3
indemnity 83.1
instruction(s) 14.3, 24.2, 27.3, 34.1, 36.1, 60.1(10), 61.2, 61.4–6, 62.1, 62.3–4, 82.1, 87.1, 91.6, C26.4, D26.4, E26.4, F26.4, X2.1
insurance 84.1–2, 85.1, 85.3–4, 86.1, 87.1–3
Key Date 14.3, 30.3
leaving Site/Working Areas 92.2
liabilities 14.1, 61.4, X15.1
liability limitation X18.1–5
low-performance damages X17.1
management responsibility F20.2
marking of Equipment, Plant and Materials 71.1

 © copyright nec 2005

method of working 35.3, A54.2, C54.2
non-corrected Defects 45.1–2
notification of compensation events 61.1–6, X2.1
notified of delegate for *Project Manager/Supervisor* 14.2
notified of replacement for *Project Manager/Supervisor* 14.4
objects of value/historical-interest 73.1
obligations 91.2
parent company guarantee X4.1
as Party to contract 11.2(11)
payment(s) 40.6, 45.1–2, 51.1, 86.1, C50.6, C52.2, D50.6, D52.2, E50.7, E52.2, F50.7, F52.2, X3.1, X3.2
performance bond X13.1
period for reply 13.3–5, 32.2
physical conditions 60.1(12), 60.2, 60.3
Plant and Materials 27.2, 41.1, 80.1, 92.2, 93.1
prevention 19.1
programme (first/revised) 11.2(1), 31.1–3, 32.1–2, 50.3, 62.2, 64.1, 64.2
proposed instruction/changed-decision 61.2
quotations 36.1, 36.2, 44.2, 62.1–5, 64.1
reaction to compensation event 63.7
repairs/replacements 82.1
reply to communication 13.3–5, 31.3, 32.2, 60.1(6), 62.3–6
responsibilities 14.1, **20–7**, C20.3, C20.4, C26.4, D20.3–4, D26.4, E20.3–4, E26.4, F20.2–4, F26.4
Risk Register revision 16.4
risks 63.6, 81.1, 84.2, 85.4
searching (for Defects) 42.1
starting/stopping work 30.1, 34.1, 91.6
subcontract contract data C26.4, D26.4, E26.4, F26.4
subcontracted work C20.3, D20.3, E20.3, F20.2, F20.3
Subcontractor 11.2(17), 26.1–3
take over of *works* 35.3, 43.4, 80.1
termination 80.1, 90.2, 90.4–5, 91.2, 91.4, 91.6–7
tests and inspections 40.2, 40.3, 40.5, 40.6, 41.1, C40.7, D40.7
title to objects and materials within Site 73.1, 73.2
title to Plant and Materials 70.1, 70.2
Works Information 11.2(19), 14.3, 63.8–9
Contractor's design **21**
acceptance 21.2, 21.3
approval from Others 27.1
and Defects 11.2(5)
and Equipment design 23.1
liability of *Contractor* 14.1
limitation of *Contractor*'s liability X15.1
management by *Contractor* F20.2
part submission 21.3
submission to *Project Manager* 21.2
use by *Employer* 22.1
and Works Information 21.1, 21.2, 60.1(1)
Contractor's employees 24.1
insurance 84.1
removing 24.2
Contractor's share C50.6, C53, C53.1–C53.4, C93.4, C93.6, D50.6, D53, D93.5–6
Contractor's share percentage C53.1, D53.5
Contracts (Rights of Third Parties) Act (1999) **Y(UK)3**
Core Group X12.2(3-5)
composition X12.1(3)
correcting
assessment of amount due 50.5, 51.3, X1.2
assumptions about compensation events 60.1(17), 61.6
Bill of Quantities mistakes B60.6, D60.6
Defects 11.2(2), 32.1, 40.4, **43**, 45.1–2, C11.2(25), D11.2(25), E11.2(25)
cost components data A11.2(22), C11.2(23), C63.15, D11.2(23), D63.15, E11.2(27), E63.15
costs
additional costs 25.3
Contractor 52.1, C40.7, D40.7
Defect(s) correction 45.1–2
due to *Adjudicator*'s/*tribunal*'s decision 51.3
Employer's risk 80.1
indemnity against 83.1, 83.2
insurance 86.1, 87.3
test/inspection repeat 40.6
see also excess costs
creditors, arrangement made with 91.1
currency of this contract 51.1, C50.6, D50.6, E50.7, F50.7, X3.1, X3.2

damage(s)
delay X7.1, X7.2
Employer's risk 80.1
insurance against 84.1
low performance X17.1
repair by *Contractor* 82.1
death insurance 84.1
decision(s)
Adjudicator or *tribunal* 51.3
assessment date 50.1
and compensation events 60.1(8), 61.1, 61.4–6
date of Completion 30.2
Project Manager 30.2, 50.1, 60.1(8), 61.1–2, 61.4–6, 63.5, 64.1, C11.2(25), D11.2(25), E11.2(25), F11.2(26)
risk reduction meeting 16.4
default(s)
Contractor's 91.2, 91.3, 91.6
Employer's 91.6
defect correction period 40.5, 43.2–4, 45.1
Defect(s)
accepting **44**, 60.1(1)(9)
Contractor's compensation X15.2
Contractor's liability 11.2(5), X15.1, X18.3
correcting 11.2(2)(6), 32.1, 40.4, **43**, 45.2, C11.2(25), D11.2(25), E11.2(25)
correction after Completion C11.2(23)
defect correction period 40.5, 43.2–4, 45.1
definition 11.2(5)
liability limitation X18.3
loss/damage/wear due to 80.1
low performance damages X17.1

 © copyright nec 2005

Defect(s) (*continued*)
notifying 11.2(6), 42.2, 43.1–2
and priced contract(s) A11.2(27), B11.2(28)
searching (for) 42.1, 60.1(10)
and tests/inspections 40.4, 40.6
uncorrected **45**, 60.1(9)
Defects Certificate 11.2(6), 43.3
and assessment of amount due 50.1
and *Contractor*'s risks 81.1–2
and *Employer*'s risks 80.1
and insurance cover 84.2
Key Performance Indicators X20.2
liability limitation X18.3
low performance damages X17.1
and retention X16.2
defects date 11.2(6), 40.5
and compensation events 61.7
searching/notifying Defects 42.1–2
testing 42.1–2
Defined Cost **52**, C11.2(29), D11.2(29), E11.2(29)(32), F11.2(29)(32)
in amount due on termination 93.1, 93.2
and compensation events 61.4, 63.1–2, 63.7, A63.10, B60.4, B63.10, B63.13, C63.11, D60.4, D63.11, X1.3, X2.1
definition(s) A11.2(22), B11.2(22), C11.2(23), D11.2(23), E11.2(23), F11.2(24)
and Fee calculation 11.2(8), 63.1, C11.2(29), D11.2(29), E11.2(29)
forecast C20.4, D20.4, E20.4, F20.4
payment on termination D93.5
payments made by *Contractor* C50.6, C52.2, C52.3, D50.6, D52.2, D52.3, E50.7, E52.2, E52.3, F50.7, F52.2, F52.3
defined terms **11**, X1.1
Delay/delay
and compensation events 62.2, 63.3, B60.5, D60.5
damages X7.1, X7.2
early warning to be given 16.1
plans to deal with 32.1
and priced contract(s) A11.2(27), B11.2(28)
and tests/inspections 40.5, 60.1(11)
delegation by *Project Manager/Supervisor* 14.2
delivery, testing and inspection before **41**
demolition 22.1, 73.2
design
Contractor's 11.2(5), **21**, 22.1, 23.1, 27.1
Employer's 80.1
of Equipment **23**
of *works* C20.3, D20.3, E20.3, F20.3
see also Contractor's design
direct fee percentage 11.2(8), 93.2
Disallowed Cost C11.2(23), D11.2(23), E11.2(23), F11.2(24)
definition(s) C11.2(25), D11.2(25), E11.2(25), F11.2(26)
discounts, deduction from prices 52.1
dismantling (during searching) 42.1
dispute resolution
unless United Kingdom Housing Grants, Construction and Regeneration Act (1996) applies **W1**
when United Kingdom Housing Grants, Construction and Regeneration Act (1996) applies **W2**

early Completion bonus X6.1
early warning **16**
effects 16.1
Key Date 16.1
responsibility 16.1, 61.5, 63.5, C11.2(25), D11.2(25), E11.2(25), F11.2(26)
Risk Register 16.1, 16.3–4
employees, of *Adjudicator* W1.2(5), W2.2(5)
Employer
access 43.4
access and use of Site 33.1, 60.1(2)
actions 10.1
adjudication W1.3(1)
advanced payment by X14.1–3
amount due to be paid by 50.2, 51.1
breach of contract by 60.1(18)
certificates issued to 13.6
co-operation 25.2, 25.3
compensation events 60.1(1-3)(5)(14)(16)(18)
Completion of *works* on termination 93.1, 93.2
Contractor's design 22.1
Contractor's parent company guarantee X4.1
dates 31.2, 33.1, 60.1(2)(3)(5)
default 91.6
Defects Certificate 43.3
facilities and services provided by 40.2, 60.1(16)
fault 80.1
hindered by *Contractor* 91.3
indemnity 83.1
insurance 84.1, 85.4, 86.1, **87**
as Party to contract 11.2(11)
payments by 51.1
payments to C11.2(23), D11.2(23), E11.2(23), F11.2(23)
performance bond X13.1
Plant and Materials 31.2, 80.1
programme 31.2, 60.1(5)
repayment of delay damages X7.1
replacement of *Project Manager/Supervisor* 14.4
responsibilities 83.2, 87.2
rights to change 63.4
risks 60.1(14), **80**, 81.1, 83.2, 85.4
take over of *works* 35.1–3, 43.4, 80.1, X6.1, X7.1, X16.1, X16.2
termination 80.1, 90.2, 91.2–3, 91.6–7, 92.1–2, 93.1–2
tests and inspections 40.2, 40.6, 60.1(16), C40.7, D40.7
title to Equipment, Plant and Materials 93.1
title to Plant and Materials **70**, 93.1
use of *works* by 11.2(2), 35.3
working with *see* co-operation
Works Information 22.1, 60.1(1)(5)(16)
Employer's design 80.1
Equipment
definition 11.2(7)
design of **23**
left on Site 72.1, 80.1
loss or damage 84.2

 © copyright nec 2005

loss, damage or wear 80.1
marking of 70.1, **71**
and programme 31.2
rates in Contract Data X1.3
removal of **72**, 92.2, 93.2
termination 92.2
title 70.1, 70.2, 92.2
use by *Employer* 92.2
see also Plant and Materials
errors, *Adjudicator* W1.3(11)
excavation materials 73.2
excess costs, *Contractor*'s share C53.2
exchange rates C50.6, D50.6, E50.7, F50.7, X3.1
experienced contractor
and compensation events 60.1(12), 60.2, 61.5, 63.5
and physical conditions on Site 60.1(12), 60.2
and prevention 19.1, 60.1(19), 91.7
extension of time for decision/quotation/reply 13.5, 62.5

facilities and services
for tests/inspections 40.2, 42.1, 60.1(16)
see also services
failure to pay in time, actions to be taken 91.4
fault of *Contractor* 61.1, 61.4, C11.2(25), D11.2(25), E11.2(25), F11.2(26)
fault of *Employer* or his design 80.1
Fee C11.2(29), D11.2(29), E11.2(29)(32), F11.2(29,32)
calculation(s) 11.2(8), 63.1, 93.2, C50.6, D50.6, E50.7, F50.7
see also payment(s)
fee percentage see direct fee percentage
float, provision in programme 31.2
forecasts 61.6, 63.1, 65.2, 93.2, C20.4, C53.3, D20.4, D53.7, E20.4, E65.3, F20.4, F65.3
form of communications 13.1
fraud, and insurance 85.2

grouping of activities A11.2(27)
guarantee(s) 91.2, X4.1

health and safety
Contractor 27.4
regulation(s) 91.3
requirements 31.2
Works Information 27.4
historical-interest objects found within Site 60.1(7), 73.1
Housing Grants, Construction and Regeneration Act (1996) **W1**, **W2**, **Y(UK)2**

identified terms **11**
illegal requirements **18**
implementing compensation events 32.1, **65**, A65.4, B65.4, C65.4, D65.4, E65.3, F65.3
impossible requirements **18**
Incentive Schedule X20.1, X20.3–5
incentive schedule X20.1
incentives, Partnering X12.4
inconsistencies **17**, 60.3, 63.8, B60.6, D60.6
indemnity **83**
information
in Activity Schedule A54.1, C54.1
in Bill of Quantities B55.1, D55.1
later recorded 65.2
in programme 31.2, 31.3, 50.3
provided to Others 25.1
publicly available 60.2
see also Site Information; Works Information
injury insurance 84.1
inspection(s) **40**, 60.1(10)(11)
before delivery **41**
Contractor's cost C40.7, D40.7
instalment repayment of advanced payment X14.3
instruction(s)
compensation events resulting 60.1(1)(4)(7), 61.1
to *Contractor* 27.3
form of communication 13.1
insurance 87.1
objects of value/interest 60.1(7)
quotation(s) to be submitted 36.1, 61.1–2, 61.5–6, 62.1, 62.3–4, 64.1, X2.1
to remove *Contractor*'s employee 24.2
on resolution of ambiguities and inconsistencies 17.1
to search (for Defects) 42.1
to stop/not start work **34**, 60.1(4), 91.6
and termination 91.6, 92.2
that change Key Date 14.3
that change Works Information 14.3, 18.1, 44.2, 60.1(1), 63.8–9
insurance
certificates 85.1, 86.1, 87.1–3
cover **84**
by *Employer* 84.1, 85.4, 86.1, **87**
policies **85**, 86.1, 87.1–3
signatories 85.1
Insurance Table 84.2
insurer(s) 85.2, 85.4, X13.1
bonds X13.1, X14.2
insurrection, risks due to 80.1
interest
assessment of 51.2, 51.3
compound calculation 51.4
on late payment 51.2
and payment certificate(s) 51.3
repaid delay damages X7.1
interest rate 51.4
interpretation
compensation event 63.8
conditions of contract 12.3
legal **12**

Key Date **30**
acceleration 36.1, A36.3, C36.3, D36.3
assessment of compensation events 63.8–9
and compensation events 60.1(4), 61.3, 62.2, A65.4, B60.5, B65.4, C65.4, D60.5, D65.4
Condition 25.3
Contractor 14.3, 30.3
definition 11.2(9)

 © copyright nec 2005

Key Date (*continued*)
early warning 16.1
notification of compensation events 61.3, 61.4
and programme 31.2
Project Manager 14.3, 60.1(4)
rights to change 63.4
Subcontractor failing to meet C11.2(23), D11.2(23), E11.2(23), F11.2(23)
Key Performance Indicators **X20**
aspect of performance X20.1
performance reports X20.2
key person 24.1

language of communications 13.1
language of this contract 13.1
late payment(s) 51.2–3, X14.2
later recorded information 65.2
Latest Index X1.1(b)
law
changes in **X2**, X2.1
and Defect(s) 11.2(5)
interpretation **12**
national legislation **Y**
release of Party from contract 91.5
tax(es) 50.2
tests and inspections 40.1
see also Acts
law of the contract 12.2
legal rights, interference with 80.1
legislation, national **Y**
liability
Contractor's design 14.1, X15.1
and insurance 84.2
limitation **X18**
reduction 83.2
liquidator for company/partnership 91.1
loss risks 80.1, 84.1
low performance
damages **X17**
see also performance impairment
lump sum(s) A11.2(30), B11.2(28), C11.2(30), C93.4, D11.2(31), D63.13

marking of Equipment, Plant and Materials **71**
materials
access to 27.2
dates when needed 31.2
testing/inspection 40.2, 41.1, 42.1, 60.1(16)
title to 73.2
see also Plant and Materials
method of measurement B60.6, D60.6
method of working 35.3, A54.2, C54.2
military power, risks due to 80.1
mistake(s)
assessment of amount due 51.3
Bill of Quantities B60.6, D60.6
multiple currencies **X3**

named person 24.1
national legislation **Y**
NEC Adjudicator's Contract W1.2(1), W2.2(1)
NEC contract 26.3, C26.4, D26.4, E26.4, F26.4
negligence 80.1
nominating body, Adjudicator W1.2(3), W2.2(3)
notification
acceptance of quotation(s) 62.3, 65.1
adjudication W1.3(1–2)(8-9), W2.3(1–2)(8–9)
ambiguities/inconsistencies 17.1
assessment of compensation event 62.3, 64.3, 65.1
communication of 13.1, 13.7
compensation events and assumptions 60.1(17), **61**
Contractor's default 91.2, 91.3
Defects 11.2(6), 42.2, 43.1–2
delegate for *Project Manager/Supervisor* 14.2
illegal/impossible requirements 18.1
implementing compensation events A65.4, B65.4, C65.4, D65.4, E65.3, F65.3
Project Manager's assessment 64.3, 65.1
Project Manager's decision(s) 63.5, 64.1
proposed instruction/changed-decision 61.2, 62.3
replacement of *Project Manager/Supervisor* 14.4
termination 90.1
tests and inspections 40.3, 41.1
notified Defects, correction 32.1, 45.2

objects of value or historical interest 60.1(7), 73.1
observing tests 40.3
open market prices 52.1
operations (in programme) 31.2, 32.1
Others (other people)
approval by 27.1
at risk reduction meeting 16.2
co-operation with 25.1
Defect(s) correction by 45.2
definition 11.2(10)
hindered by *Contractor* 91.3
Plant and Materials supplied by 80.1
and programme 31.2, 60.1(5)
working with see co-operation
and Works Information 60.1(5)
Own Contract, definition X12.1(2)

parent company guarantee **X4**
Parties/Party
Adjudicator's communications W1.3(6), W2.3(6)
Adjudicator's instructions W1.3(5), W2.3(4)
agreement 12.4
and assessment date 50.1
to contract 11.2(11)
indemnity liability 83.1, 83.2
insurance 84.2, 85.3
prevention 19.1
to subcontract 26.3
and termination 90.1, 91.1, 91.5
third party rights Y3.1
tribunal reviews W1.4(1–4), W2.4(1–3)(5)
Partnering **X12**
actions X12.2
change of Partnering information X12.3(6)
Core Group X12.2(3-5)
defined terms X12.1

 © copyright nec 2005

early warning X12.3(3)
incentives X12.4
Key Performance Indicators X12.4(1–2)
Own Contract X12.1(2)
Partnering Information X12.1(4), X12.3(1)
representatives X12.2(2)
subcontracting X12.3(9)
timetable X12.3(7)
payment(s) **51**
advanced 93.1, X14.1–3
certified 50.5, 51.1–3, 90.4
conditional on success of test or inspection 40.5
Contractor 40.6, 45.2, 51.1, 86.1, C50.6, C52.2, D50.6, D52.2, E50.7, E52.2, F50.7, F52.2, X3.1–2
dates Y2.2
Employer's insurance 87.3
Housing Grants, Construction and Regeneration Act (1996) Y2.2–3
interest 51.4
late 51.2–3, X14.2
and marking of Equipment, Plant and Materials 71.1
not made 91.4
notice of intention Y2.3
taxes 50.2
on termination 93.1, 93.2, A93.3, C93.4, C93.6, D93.5–6
time to be made 51.2, X14.2
withholding Y2.3
payment certificate 50.5, 51.1, 51.3, 91.4
payment clauses **50**
people **24**
rates in Contract Data X1.3
see also employees
performance, *see also* Key Performance Indicators
performance bond **X13**, X13.1
performance impairment
early warning of possibility 16.1
see also low performance
performance reports X20.2
performance suspension Y2.4
period for reply 13.3, 13.4, 32.2, 36.2
extension agreed 13.5
personal injury 84.1
see also insurance
physical conditions of Site 60.1(12), 60.2, 60.3
Plant and Materials 11.2(12)(17)
access to 27.2, 60.1(2)
dates when needed 31.2
loss of or damage to 80.1, 82.1, 84.1
not used to Provide the Works C11.2(25), D11.2(25), E11.2(25)
removal of 92.2
replacement 82.1, 84.1
and termination 92.1, 93.1
testing/inspection before delivery 41.1
title 70.1, 70.2, 92.1, 93.1
wastage C11.2(25), D11.2(25), E11.2(25)
see also Equipment
prevention **19**
Price Adjustment Factor(s) **X1**
assessment of compensation events X1.3
calculation of X1.2
definition X1.1(c)
price adjustment use X1.4, X1.5
Price for Work Done to Date
adjustment for inflation X1.4, X1.5
and assessment of amount due 50.2, 50.3, 93.2, C93.4, D93.5
and assessment of *Contractor*'s share C53.1–4, D53.5–8
definition(s) A11.2(27), B11.2(28), C11.2(29), D11.2(29), E11.2(29), F11.2(29)
and retention X16.1
Prices/prices/*prices*
and acceleration 36.1, A36.3, B36.3, C36.3, D36.3
and Activity Schedule A11.2(20)(27), A54.3, A63.12, C11.2(30), C54.3, C63.12
and Bill of Quantities B11.2(31), B60.4, B60.6, B63.13, D11.2(31), D60.4, D60.6, D63.13
changes to 63.1–2
and compensation events 61.4, 62.2, 63.1–2, A63.10, A65.4, B60.4, B60.6, B63.10, B63.13, B65.4, C63.12, C65.4, D60.4, D60.6, D63.13, D65.4, E65.3, F65.3, X2.1
competitively tendered/open market 52.1
Contractor's share C53.1–C53.4, D53.5
definition A11.2(30), B11.2(31), C11.2(30), D11.2(31), E11.2(32), F11.2(32)
early warning of possible increase 16.1
if Defect(s) uncorrected/accepted 44.2
notification of compensation events 61.3
rights to change 63.4
and termination 93.2, C93.4, D93.5
procedures, termination 90.2, 90.3, **92**
proceedings
Employer's risk 80.1
indemnity against 83.1, 83.2
procurement procedure C11.2(25), D11.2(25), E11.2(25), F11.2(26)
programme 11.2(1), **31**, 50.3, 62.2, 64.1
accelerated 36.1, A36.3, B36.3, C36.3, D36.3, E36.4, F36.4
dates to be shown 31.2, 32.1, A31.4, C31.4
revised **32**, 36.1, 62.2, 64.2, A36.3, B36.3, C36.3, D36.3, E36.4, F36.4
Project Manager
acceleration 36.1, A36.3, B36.3, C36.3, D36.3, E36.4, F36.4
acceptance of communication/proposal/submission 13.4, 14.1, 15.1, 21.2, 31.3, 85.1
acceptance of contract data for subcontract C26.4, D26.4, E26.4, F26.4
acceptance of quotation 44.2, 60.1(9), 62.3, 65.1, A36.3, B36.3, C36.3, D36.3, E36.4, F36.4
access 27.2
Activity Schedule A54.2, C54.2
addition to *working areas* 15.1
Adjudicator's reviews/revisions W1.3(5), W2.3(4)
advanced payment bond X14.2
and ambiguities/inconsistencies 17.1
assessment of amount due 50.1, 50.4, 50.5

 © copyright nec 2005

Project Manager (*continued*)
assessment of compensation events 61.6, 63.9, **64**, A63.10, B63.10, B63.13, C63.15, D63.13, D63.15, E63.15
assessment of *Contractor*'s share C53.1, C53.3–4, C93.4, C93.6, D53.5, D53.7–8, D93.5–6
assessment date decided by 50.1
assessment of Defect(s) correction cost 45.2
assessment of test/inspection cost 40.6
assumptions about compensation events 60.1(17), 61.6
certificates 13.6, 30.2, 50.5, 51.1–3, 60.1(15), 82.1, 90.1, 90.3–4, C93.4, D93.5
co-operation 10.1, 16.3
compensation events 60.1(1)(4)(6)(7)(8)(9)(15)(17), 61.1–6, A63.10, B63.10, C63.15, D63.15, E63.15, F20.5
Completion 30.2
Contractor's advice C20.3, D20.3, E20.3, F20.3
Contractor's default 91.2, 91.3
Contractor's design 11.2(5), 21.2
Contractor's employees 24.1, 24.2
correction of mistakes in Bill of Quantities B60.6, D60.6
date of Completion 30.2
decisions 30.2, 50.1, 60.1(8), 61.1–2, 61.4–6, 63.5, 64.1, C11.2(25), D11.2(25), E11.2(25), F11.2(26)
Defect(s) acceptance/non-correction 44.1, 44.2
delegation 14.2
Disallowed Cost decision C11.2(25), D11.2(25), E11.2(25), F11.2(26)
duties required 10.1, **14**
early warning 16.1, 16.2, 16.4, 61.5, 63.5
Equipment design 23.1
forecasts of Defined Cost C20.4, D20.4, E20.4, F20.4
illegal/impossible requirements 18.1
implementing compensation events 65.1, E65.3, F65.3
inspection of *Contractor*'s accounts and records C52.3, D52.3, E52.3, F52.3
instruction(s) 14.3, 17.1, 18.1, 24.2, 27.3, 34.1, 44.2, 60.1(1)(4)(7), 61.1–2, 61.4, 62.1, 62.3–4, 82.1, 87.1, 91.6, C26.4, D26.4, E26.4, F26.4, X2.1
insurance 85.1, 87.1
interest on payments 51.3
Key Date 14.3, 60.1(4)
Key Performance Indicators X20.3
moving of Equipment, Plant and Materials 70.2, 72.1
notification of compensation events 61.1–6, X2.1
notification of *Contractor*'s default 91.2, 91.3
notification of implementing compensation events E65.3, F65.3
objects of value/historical-interest 60.1(7), 73.1
performance bond X13.1
prevention 19.1
programme 11.2(1), 31.1, 31.3, 32.2, 50.3
quotations for acceleration 36.1–2, 60.1(9), A36.3, B36.3, C36.3, D36.3
quotations for compensation events 62.1, 62.3–5
replacement of 14.4
reply to communication 13.3–5, 60.1(6), 62.3, 62.5–6
request to remove resources from Working Areas C11.2(25), D11.2(25), E11.2(25)
Risk Register revision 16.4
start/stop of work 34.1, 60.1(4), 91.6
subcontract conditions 26.3
subcontracting C26.4
Subcontractor acceptance 26.2, 91.2
take over of *works* 35.3, 43.4, 60.1(15)
and termination 90.1, 90.3, 90.4, 91.2
tests and inspections C40.7, D40.7
Works Information 14.3, 18.1, 44.2, 60.1(1)
property, loss or damage 84.1
proposal(s)
acceleration 36.1
to add to Working Areas 15.1
Defect(s) acceptance/non-correction 44.1
form of communication 13.1
risk reduction meeting 16.3, 16.4
proposed instruction or changed decision 61.2, 62.3
Provide the Works **20**
and addition to Working Areas 15.1
Contractor's responsibilities 11.2(7), 14.1, 20.1, 23.1, 26.3, C20.3–4, C26.4, D20.3–4, D26.4, E20.3–4, E26.4, F20.3–5, F26.4
and Equipment 11.2(7), 23.1
meaning of term 11.2(13)
Plant and Materials not used C11.2(25), D11.2(25), E11.2(25)
and programme 31.2
resources not used C11.2(25), D11.2(25), E11.2(25)
and Works Information 11.2(19), 20.1, C11.2(25), D11.2(25), E11.2(25)
publicly available information 60.2

quantities B11.2(28)(31), B60.4–7, D11.2(31), D60.4–7
see also bill of quantities; Bill of Quantities
quotations
for acceleration 36.1–2, 60.1(9), A36.3, B36.3, C36.3, D36.3, E36.4, F36.4
acceptance of 36.1, 44.2, 60.1(9), 62.3, 65.1, A36.3, B36.3, C36.3, D36.3
for compensation events 61.4, 61.5, **62**, 64.1, 64.3–4, 65.1, A11.2(22), C11.2(25), D11.2(25), E11.2(25), F11.2(26), X2.1
if Defect(s) accepted 44.2, 60.1(9)
for proposed instruction/changed-decision 61.2
revised 62.3, 62.4, 64.1

radioactive contamination 80.1
rates
in assessing compensation events A63.14, B63.13, D63.13
in Bill of Quantities B11.2(28)(31), B60.4, D11.2(25,31), D60.4, D63.13
in Contract Data X1.3, X6.1, X7.1
rebates, deduction from prices 52.1

 © copyright nec 2005

rebellion, risks due to 80.1
receipt of communications 13.1
receiver over assets 91.1
records
Contractor's C11.2(25), C52.2–3, D11.2(25), D52.2–3, E11.2(25), E52.2–3, F11.2(26), F52.2–3
form of communication 13.1
risk reduction meeting 16.4
Risk Register revision 16.4
re-covering (during searching) 42.1
re-erecting (during searching) 42.1
referral, adjudication W1.3(1–3), W2.3(1–3)
removing
Contractor's employee 24.2
Equipment **72**, 92.2, 93.2
Plant and Materials 92.2
resources C11.2(25), D11.2(25), E11.2(25)
repairs **82**
replacement(s)
of *Adjudicator* W1.2(4), W2.2(4)
Contractor's employee 24.1
Equipment 80.1, 84.1
Plant and Materials 82.1, 84.1
Project Manager/Supervisor 14.4
reply
form of communication 13.1
time allowed 13.3–5, 31.3, 32.2, 36.2, 60.1(6), 62.5–6
resignation of *Adjudicator* W1.2(3), W2.2(3)
resources not used to Provide the Works C11.2(25), D11.2(25), E11.2(25)
responsibilities
Contractor 14.1, **20–7**, C20.3, C20.4, C26.4, D20.3–4, D26.4, E20.3–4, E26.4, F20.2–4, F26.4
Contractor's 11.2(7), 14.1, **20**, 83.2
Employer's 83.2, 87.2
retention 50.3, 93.1, **X16**
retention free amount X16.1
retention percentage X16.1
reviews, *tribunal* W1.4(1–6), W2.4(1–5)
revised Activity Schedule A54.2, A54.3, C54.2, C54.3
revised programme 32.1–2, 36.1, 64.2, A36.3, B36.3, C36.3, D36.3, E36.4, F36.4
revised quotation 62.3, 62.4, 64.1
revolution, risks due to 80.1
Rights of Third Parties (Contracts) Act **Y(UK)3**
riot risks 80.1
risk allowances 63.6
risk reduction meeting
purpose 16.3
records 16.4
who attends 16.2
Risk Register
definition 11.2(14)
early warning 16.1, 16.3–4
Project Manager revision 16.4
records revision 16.4
risks
Contractor's 63.6, **81**, 83.2, 84.2, 85.4
Employer's 60.1(14), **80**, 81.1, 83.2, 85.4

safety *see* health and safety
samples 40.2, 42.1, 60.1(16)
savings, *Contractor*'s share C53.2
Schedule of Cost Components C11.2(23), D11.2(23), E11.2(23)
Shorter Schedule A11.2(22), B11.2(22), C63.15, D63.15, E63.15
Schedule of Partners X12.1(1)
searching for Defects 42.1, 60.1(10)
section of the *works* X5.1
sectional Completion **X5**
services 11.2(13)
see also facilities and services
share ranges C53.1, D53.5
Shorter Schedule of Cost Components A11.2(22), B11.2(22), C63.15, D63.15, E63.15
Site
access to 31.2, 60.1(2)
access and use of 33.1
definition 11.2(15)
historical-interest/valuable objects found 60.1(7), 73.1
information describing 11.2(16)
location X2.1
physical conditions 60.1(12), 60.2, 60.3
removal of Equipment, Plant and Materials 72.1, 92.2
and risks 80.1
starting of work 30.1
termination 92.2
visual inspection 60.2
Site Information 11.2(16), 60.2
and Activity Schedule A54.1, A54.2, C54.1
ambiguities/inconsistencies 60.3
and Bill of Quantities B55.1, D55.1
starting date 31.2, 32.2, 50.1, 81.1, 84.2, 87.1, C20.4, D20.4, E20.4, F20.4
dispute resolution W1.2(1), W2.2(1)
insurance 85.1
start(ing) work **30**
instruction(s) given not to 34.1, 60.1(4), 91.6
stopping work 34.1, 60.1(4), 91.6
strikes, risks due to 80.1
subcontract
adjudication W1.3(4), W2.3(3)
contract data C26.4, D26.4, E26.4, F26.4
subcontracted fee percentage 11.2(8)
subcontracting **26**, C20.3, C26.4, D20.3, D26.4, E20.3, E26.4, F20.2, F20.3, F26.4
assignment of benefits 92.2
conditions 26.3
Contractor's management responsibility F20.2
Defined Cost A11.2(22), C11.2(23), D11.2(23), E11.2(23), F11.2(26)
Subcontractor
acceptance by *Project Manager* 26.2, 91.2
appointment 26.2, 26.3, 91.2
compensation event assessments C52.2, D11.2(25), D52.2, E11.2(25), E52.2, F11.2(26), F52.2

 © copyright nec 2005

Subcontractor (*continued*)
definition 11.2(17)
payments made C11.2(23)(25), D11.2(23)(25), E11.2(23)(25), F11.2(24)(26)
submission(s)
addition to Working Areas 15.1
communications 13.1, 13.4
Contractor's design 21.2
Equipment design 23.1
forecasts of total Defined Cost C20.4, D20.4, E20.4, F20.4
insurance policies/certificates 85.1, 86.1, 87.1, 87.3
non-acceptance of 13.8, 15.1, 21.2, 23.1, 26.2, 26.3
programme (first/revised) 31.2, 32.2, 50.3, 64.1, 64.2, A31.4, C31.4
quotations 36.1, 36.2, 44.2, 61.4, 62.1, 62.3–6, 64.1, X2.1
Subcontractor/subcontract conditions/contract data 26.2, 26.3, C26.4, D26.4, E26.4, F26.4
subrogation rights 85.2
Supervisor
access 27.2
Adjudicator's reviews/revisions W1.3(5), W2.3(4)
certificates issued by 13.6, 43.3, 50.1
and compensation events 60.1(6)(8)(10)(11), 61.1
Defects notified by 11.2(6), 42.2, 43.1, 43.3
delegation 14.2
duties required 10.1, **14**
instructions given by 27.3, 61.1
marking of Equipment, Plant and Materials 70.1, 71.1
replacement of 14.4
reply to communication 13.3, 14.1, 60.1(6)
searching (for Defects) 42.1, 60.1(10)
tests and inspections 40.3, 40.5, 41.1, 60.1(11)
suppliers 11.2(10)(17)

take over of *works* **35**, 43.4, 60.1(15), 80.1, X6.1, X7.1, X16.1, X16.2
taxes 52.1
payment 50.2
termination **90**
amount due on 90.2, 90.4, 93.1, 93.2, A93.3
certification of 90.1, 90.3, 90.5, C93.4, D93.5
and insurance cover 84.2
loss/damage risks 80.1
payment on 93.1, 93.2, A93.3, C93.4, D93.5
procedures 90.2, 90.3, **92**
reasons for 90.2, **91**
Termination Table 90.2, 92.2, 93.2
tests/testing **40**
before delivery **41**
compensation events resulting 60.1(10)(11)(16)
Contractor's cost C40.7, D40.7
searching/notifying Defects 42.1–2
third party rights, contracts Act Y3.1
time, *see also* Completion; Key Date; start(ing) work
time allowed
certification of Completion 30.2
certification of payment 50.5, 51.1, 90.4
certification of take over 35.3
compensation events 61.1, 61.3, 61.4, 64.3–4
Defect(s) correction 43.2
extension(s) agreed 13.5, 62.5–6
payment of amount due 51.2, 91.4
programme submission/acceptance 31.1, 31.3, 32.2
Project Manager not replying 64.4
reply to communications 13.3, 13.5, 31.3, 32.2, 60.1(6)
start/restart of work 91.6
time clauses **30**
time risk allowances 31.2, 63.6
timings, adjudication W1.3(8)(10-11), W2.3(2)(8)(11–12)
title **70**
Equipment, Plant and Materials 92.1, 92.2
materials from excavation and demolition 73.2
object of value or historical interest 73.1
Plant and Materials **70**
total of the Defined Cost, payment on termination D93.5
Total of the Prices
Contractor's share D53.5–8
identified and defined terms D11.2(33)
total of the Prices
Activity Schedule A54.3, C54.3
adjustment for inflation X1.5
compensation events B60.4, D60.4
Contractor's share C53.1, C53.3
early warning of possible increase 16.1
identified and defined terms 11.2(27)
payment on termination 93.2, C93.4, C93.4
tribunal
Adjudicator W1.3(2)(10), W2.3(11)
arbitration W1.4(5), W2.4(4)
costs D11.2(25)
decisions 51.3
proceedings C11.2(25), D11.2(25), E11.2(25)
reviews W1.4(1–6), W2.4(1–5)
trust and co-operation 10.1, 26.3, X12.3(1)

uncorrected Defects **45**, 60.1(9)
uncovering work (during search) 42.1
undecided disputes, *Adjudicator* W1.2(4), W2.2(4)
United Kingdom Housing Grants, Construction and Regeneration Act (1996) **W1**, **W2**
use of *works* by *Employer* 11.2(2), 35.3
usurped power, risks due to 80.1
utilisation of resources C11.2(25), D11.2(25), E11.2(25)

valuable objects found within Site 60.1(7), 73.1
visual inspection of Site 60.2

waiver by insurers 85.2
war 80.1
wastage of Plant and Materials C11.2(25), D11.2(25), E11.2(25)
watching tests 40.3
wear to *works*, *Employer*'s risk 80.1

 © copyright nec 2005

weather conditions 60.1(12)
weather data/weather measurement 60.1(13)
whole of the *works* X5.1
  Completion 32.2, 50.1, 60.1(13), 93.1–2, C20.4, C53.3, D20.4, D53.7, E20.4, F20.4, X1.2, X16.1
  Defined Cost C20.4, D20.4, E20.4, F20.4
winding-up order 91.1
witnesses, *tribunal* proceedings W1.4(6), W2.4(5)
Working Areas (*working areas*) 11.2(18)
  addition **15**
  Equipment, Plant and Materials outside 71.1
  Equipment, Plant and Materials removed from 92.2
  Equipment, Plant and Materials within 41.1, 93.1
  Plant and Materials outside 70.1, 70.2, 71.1
  Plant and Materials within 41.1, 70.2, 93.1
  resources removed from C11.2(25), D11.2(25), E11.2(25)
  sharing with Others 25.1
working with *Employer*/Others *see* co-operation
works
  Completion of whole 11.2(13), 32.2, 50.1, 60.1(13), 92.1, 93.2, C20.4, C53.3, D20.4, D53.7, E20.4, F20.4, X1.2, X16.1
  Contractor's design 21.1, 22.1, X15.1
  Contractor's management responsibility for design, construction and installation F20.2
  and Defects 11.2(5)
  design C20.3, D20.3, E20.3, F20.3
  early warning of possible performance impairment 16.1
  Equipment left in 72.1
  information specifying/describing 11.2(19), 25.1
  loss of or damage to 80.1, 82.1, 84.1
  part provided by Subcontractor 11.2(17)
  Plant and Materials included 11.2(12)(17)
  section X5.1
  taking over **35**, 43.4, 60.1(15), 80.1, X6.1, X7.1, X16.1, X16.2
Works Information
  acceptance/procurement procedure C11.2(25), D11.2(25), E11.2(25), F11.2(26)
  accounts and records C52.2, D52.2, E52.2, F52.2
  and Activity Schedule A54.1, C54.1
  advanced payment bond X14.2
  assessing compensation events A63.10, B63.10
  and Bill of Quantities B55.1, D55.1
  change(s) 14.3, 18.1, 44.1–2, 60.1(1), 63.8–9
  and compensation events 60.1(1)(16), 63.8–9
  Completion Date 11.2(2)
  conditions stated in 25.1, 60.1(5)
  Contractor's accounts and records C52.2, D52.2, E52.2, F52.2
  Contractor's design 21.1–2, 22.1, 60.1(1), X15.1
  Contractor's parent company guarantee X4.1
  and Defects 44.1, 44.2
  definition 11.2(19)
  Equipment 11.2(7)
  health and safety 27.4
  instructions that change 14.3, 18.1, 44.2, 60.1(1), 63.8–9
  marking of Equipment, Plant and Materials 71.1
  part of *works* not in accordance with 11.2(5)
  performance bond X13.1
  and programme 31.2
  and Providing the Works 11.2(19), 20.1, C11.2(25), D11.2(25), E11.2(25)
  subcontract work F20.2
  take over of *works* 35.2
  tests and inspections 40.1–2, 41.1, 42.1, 60.1(16)
  title to materials 73.2
  uncorrected Defects 45.1–2

 © copyright nec 2005